Mainsail Trimming

An Illustrated Guide

FELIX MARKS

Mainsail Trimming

An Illustrated Guide

FELIX MARKS

With Photography by Neil Hinds and Felix Marks

John Wiley & Sons, Ltd

Published by John Wiley & Sons Ltd, The Atrium, Southern Gate, Chichester, West Sussex PO19 8SQ, England
Telephone (+44) 1243 779777

Email (for orders and customer service enquiries): cs-books@wiley.co.uk
Visit our Home Page on www.wiley.com
Reprinted Dec 2009

Other Wiley Editorial Offices

John Wiley & Sons Inc., 111 River Street, Hoboken, NJ 07030, USA

Jossey-Bass, 989 Market Street, San Francisco, CA 94103-1741, USA

Wiley-VCH Verlag GmbH, Boschstr. 12, D-69469 Weinheim, Germany

John Wiley & Sons Australia Ltd, 42 McDougall Street, Milton, Queensland 4064, Australia

John Wiley & Sons (Asia) Pte Ltd, 2 Clementi Loop #02-01, Jin Xing Distripark, Singapore 129809

John Wiley & Sons Canada Ltd, 6045 Freemont Blvd. Mississauga, Ontario, L5R 4J3

Wiley also publishes its books in a variety of electronic formats. Some content that appears in print may not be available in electronic books.

Anniversary Logo Design: Richard J. Pacifico

Library of Congress Cataloguing in Publication Data

Marks, Felix.
Mainsail trimming: an illustrated guide / Felix Marks ; with photography by Neil Hinds and Felix Marks.
p. cm.
Includes index.
ISBN 978-0-470-51650-8 (pbk.: alk. paper)
1. Sails. 2. Sailing. I. Title.
VM532.M234 2007
623.88'22–dc22 2007011027

British Library Cataloguing in Publication Data

A catalogue record for this book is available from the British Library

ISBN 978-0-470-51650-8 (PB)

Typeset in 10/15 Futura by Thomson Digital, India
Printed in Singapore by Markono Print Media Pte Ltd

Photographs by Neil Hinds and Felix Marks
Illustrations by Felix Marks

Acknowledgements

The author gratefully acknowledges the help of Neil Hinds for photography. Melissa Collins, Lindy Baker, Lucy McInnes, Conrad Johnston, Martin White and Judith Collins are also thanked for their help in developing this book.

Contents

Acknowledgements v

Foreword ix

1 Creating Lift and Avoiding Drag 1
- Creating lift with airfoils 2
- Avoiding drag with airfoils 5
- Overall mainsail trimming: goals and means 5
- Sail shape: belly (depth) 8
- Sail shape: leech (twist) 9

2 The Mainsail and its Controls 11
- The Mainsheet 12
- Traveler 16
- Vang 22
- Halyard 26
- Cunningham 26
- Outhaul 28
- Backstay 30

3 The Complete Shapes 35
- The complete upwind shapes 36
- Reading the telltales 38

4 How to Sail Fast Upwind 43
- Gears 45
- Pointing only comes with speed! 47

Keeping speed 49
Controlling heel 50
Using 'feel' 54
Using target boat speeds 55
Fast maneuvers 57

5 Offwind Trim 65
Reaching 66
Running 67
Jibing 69

6 Hoisting, Dropping, and Reefing 73
Hoisting 74
Dropping 75
Reefing 76

Conclusion 79

Appendix A – Traveler Track 81

Appendix B – Trim Table Template 83

Appendix C – Points of Sail 85

Glossary 87

Index 93

Foreword

So You Want to be a Mainsail Trimmer

Mainsail trimming is one of the most physically and mentally demanding roles on a boat. It's also one of the most important. It requires a deep and subtle appreciation of sail shapes and the controls used to achieve them. Most significantly, the role requires an obsession with acquiring and retaining speed. This book is here to help and has been written in a groundbreaking way. The approach we have taken is to explain everything that you need to know without blinding you with science. Sailing terminology has been reined in as much as possible and only information relating to mainsail trimming has been included. We have not tried to explain tactics, navigation, strategy, jib trimming, helming, spinnaker work, or any of the many other subjects you simply don't need to understand in order to be able to trim the mainsail well.

Mainsail Trimming is About Creating Boat Speed...fast!

As you baulk at the prospect of taking on such a crucial role, rest assured that with the information contained within this book, you will soon be trimming with confidence and ability. The guide is full of annotated photographs and diagrams that show you exactly what you're trying to achieve. We have also explained the many 'secrets' that help others stand out from the crowd. Once this guide is in your head, you will be trimming like a pro.

You're Part of a Team!

On smaller boats, the mainsail trimmer also steers. On larger boats, the mainsail trimmer is part of a team where there are many other responsibilities and activities, including jib trimming, helming, spinnaker trim, mast, pit, foredeck and tactics. The smaller the boat, the more these responsibilities are combined into one person. A key to doing well as a mainsail trimmer is to understand how you relate to the other core functions on the boat, especially the helm. This book highlights when and how you can expect others to act.

Creating lift and avoiding drag

Creating Lift with Airfoils

Avoiding Drag with Airfoils

Overall Mainsail Trimming: Goals and Means

Sail Shape: Belly (Depth)

Sail Shape: Leech (Twist)

This section explains the basic principles of how sails work, including airfoils, lift and drag, sail depth and twist, and overall trimming goals.

Creating Lift with Airfoils

Before you start looking at sail shape, you must first understand a little of how sails work.

- When a boat is sailing, its sails are its engines. Sails use wind energy to create driving force. This force is harnessed to move a boat through (and sometimes over) the water.
- Sails can be used as airfoils or air dams. When the mainsail is used for upwind sailing it's used as an airfoil. Downwind sailing means the sail is used as an air dam. The crossover point is, approximately, when you're sailing on a broad reach (see Appendix C).
- Airfoils are special shapes that create lift and drag. Lift is the useful force that we use to make the boat go forward. Drag represents the forces that slow the boat down. Good sail trimming is about maximizing lift and minimizing drag.

Having promised not to blind you with science, I will however, be giving you a little bit of theory you can't do without if you're going to understand lift and drag. It's pretty straightforward though!

- Higher pressure air tries to move towards lower pressure air. Anything between high and low pressure experiences a force towards the lower pressure too. For example, when you burst a balloon, the higher pressure air inside the balloon escapes. As it does so, the outside of the balloon is blown away – towards lower pressure air.
- The same principle applies to airfoils such as airplane wings or sails.
- Airfoils are used to create a pressure difference. This pressure difference generates lift. Lift is a force that we harness in sailing to make a boat move forwards. In aviation, this force is harnessed to elevate aircraft.

Airfoil: Airplane Wing

The engines on an airplane move the plane forward. This causes air to flow over the wings. There is a relative difference between the speeds at which the same amount of air travels over the wing compared to under. This is because the air traveling over the wing has further to travel. This relative difference in speed causes a pressure difference that lifts the wing, and with it the airplane.

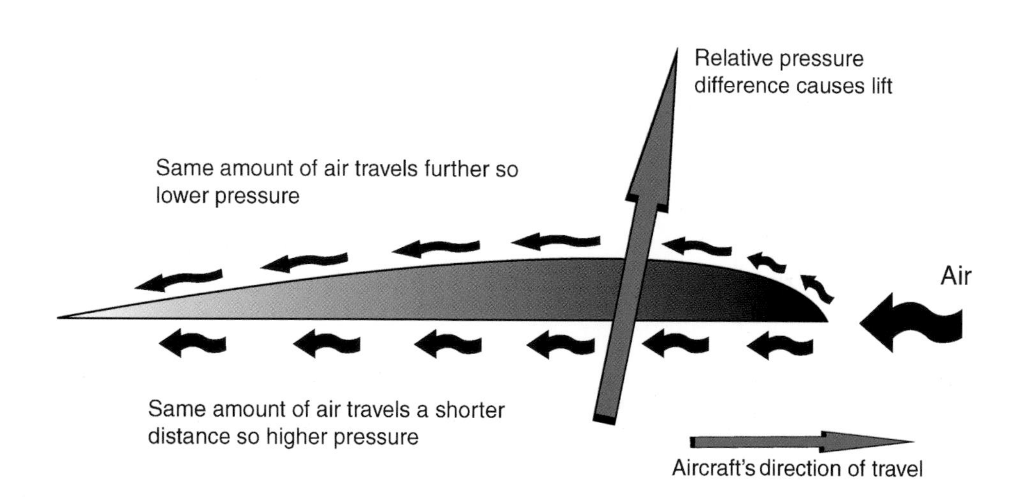

Airfoil: Sail

As the wind flows around a sail, the relative difference in the speeds that the air moves causes a pressure difference that lifts the sail forward, and hence the boat too. Note that an airplane's wing is horizontal and a yacht's sail is vertical which is why lift on a yacht forces the boat forward, and lift on an airplane forces the airplane up.

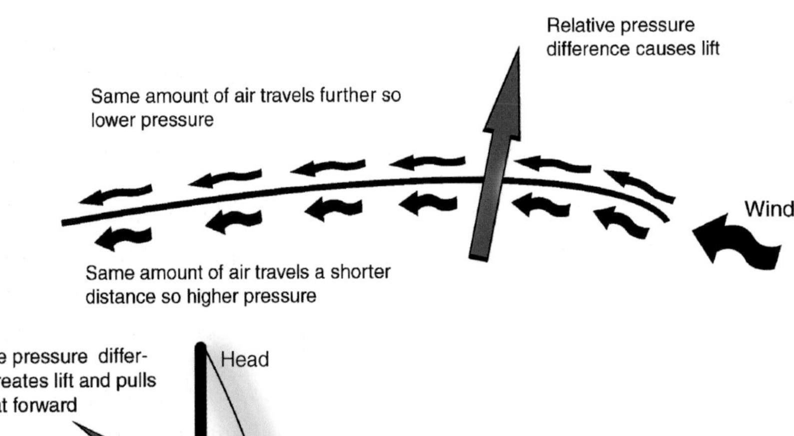

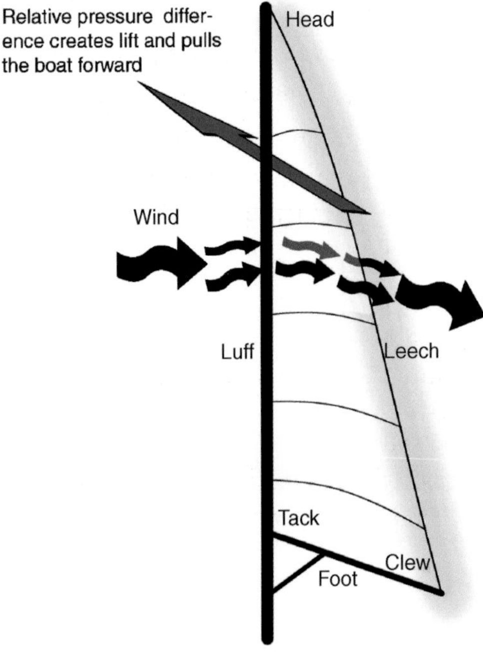

Avoiding Drag with Airfoils

Now that you understand lift, let's look at drag.

- Drag sounds bad – and it is bad. Too much drag will cause the boat to become less controllable and as a result, it will go slower.
- Drag is created as a side effect of lift and it primarily causes a boat to be knocked over. This is called heeling. While a bit of heel is often good, too much slows the boat down. The more a boat heels, the more the helmsman will have to compensate with the rudder. The more the helmsman uses the rudder, the slower the boat goes because it ends up acting like a brake.
- On a windy day out on the water, many boats will have much too much heel, i.e. over 20 degrees. Many will have over 60 degrees of heel and some will almost be flattened. This is bad sailing and is entirely avoidable. As a mainsail trimmer, you are primarily responsible for balancing lift and drag.
- When a boat is sailing well, it is because we have created the right balance in the sails between lift and drag. The following sections explain the fundamentals of mainsail shape. It is essential for you to understand what mainsail shapes there are and how you control them. With practice, you will be able to see immediately whether the sail is the right shape for the conditions.

Overall Mainsail Trimming: Goals and Means

Now that you've covered the basics of lift and drag, it's time to understand the fundamentals of mainsail trim.

Goals:

- To develop as much power as possible to go as fast as possible.
- To control drag so the boat is manageable and going fast.

- To enable the helmsman to steer in the right direction.

Therefore, the correct sail shape is the best compromise between lift and drag.

Means:

The power in the sails is controlled in two ways: rig set-up (i.e. shroud tension and backstay tension) and mainsail trim (sheet, halyard, outhaul, vang and cunningham).

1. Rig Set-up

- Rig tension is controlled by the skipper, though the job of making changes to the rig may have been delegated – perhaps even to you! The purpose of changing the rig tension is to control the overall power that the sails can develop.
- You're in big trouble if your rig's out by very much. Racing in light winds with a rig set-up for big winds will see you wallowing at the back of the fleet. Your sails won't be powerful enough. A rig set-up for light winds but sailed in heavy winds will make your day very hard indeed. Your sails will be too powerful and so it will be difficult to maintain control.
- When you're racing, the rules tend to limit the changes you're allowed to make to the rig. Changes to the rig can only be made before the preparatory period of a race.
- Importantly though, the backstay can be changed throughout a race.

What does this mean to you as the mainsail trimmer?

- However well you trim your mainsail, you cannot develop more lift from the sails than is available from the rig. Since your job is to help make the boat sail well, you need to be aware that poor performance may be coming from somewhere else on the boat. If the boat is going slower than expected, and perhaps the blame is being leveled at you or the jib trimmer, you need to ask the following question: is the rig too tight for these conditions?

- For a rig set tight, however much you ease the backstay and outhaul, you will never be able to get more lift from the sails beyond the limit set by tight rig.
- Conversely, for a rig set too loose, however hard you tension the backstay and outhaul, you will never be able to reduce the drag from the sails below the limit set by the loose rig.

Although the difference between a loose and tight rig is hard to see, the effect is significant in terms of the overall power the sail generates. A quick and easy way to check the rig's tension is its effect on the forestay.

Firstly, make sure the backstay is off completely. Then, grab the forestay and push against it.

- A tight forestay will hardly yield and indicates a tight rig.
- A loose forestay will move appreciably and indicates a loose rig.

Apart from the effects of rig tuning, the subject of rig tuning itself is not discussed in this book.

As a member of a team of sailors, you should ask the skipper before you start sailing what the rig tension is being set to. After all, it affects you as well as the rest of the boat. Setting the correct rig tension heavily influences the course of your day on the water.

2. Mainsail Trim

The other means of controlling power in the mainsail is the subject of the rest of the book.

Sail Shape: Belly (Depth)

Sail depth, or belly size, controls the power in the sail and hence how much lift and drag is created.

- The sail is at its most powerful, and creating most lift, when its belly is at its biggest setting. However, sailing with a big belly in high winds creates too much drag, which causes the boat to heel over, become uncontrollable and slow down.

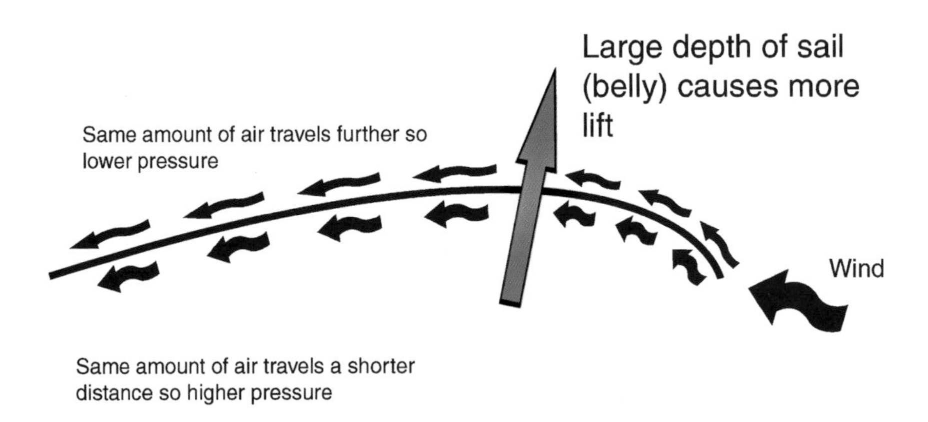

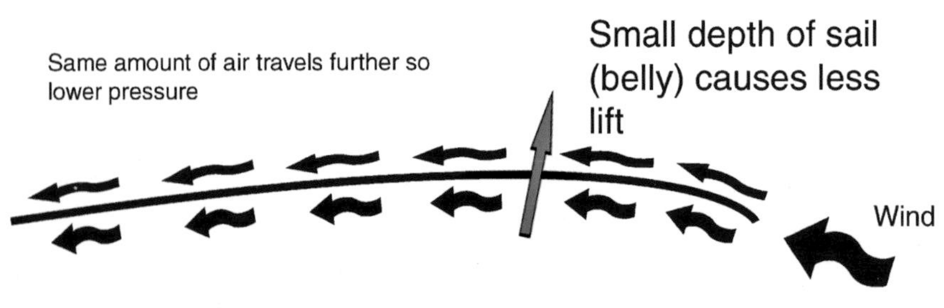

- A big belly is suitable for light winds and a small belly is suitable for strong winds. You reduce the size of the belly as the winds get stronger but you only start reducing the belly as the forces of drag affect your performance and start to slow you down. As the wind drops in strength, you increase the belly again.

A deeper sail (bigger belly) produces more lift and drag. A sail with a smaller belly creates less lift and drag. Although the lift is welcome, in higher winds the drag over powers the boat and makes it uncontrollable and slow. Therefore, we vary the depth of the sail according to the wind strength.

Sail Shape: Leech (Twist)

The shape of the mainsail's leech is referred to as twist. If the leech at the top of the sail is angled further out from the boat than at the foot, we say it is open and therefore has twist. The amount of twist simply refers to how open, or closed, the top of the leech is.

- Understanding how much twist is in the sail is easy. Look at the angle the boom is pointing from the centreline of the boat. Now, at the same time, run your eyes up the leech to the top batten. The angle between the top batten and the boom is the amount of twist in the sail.

Twist is affected by the mainsheet, traveler and vang tensions. The tighter the mainsheet, the less twist, and hence the more closed the top of the leech becomes. Similarly, the tighter the vang, the more closed the top of the leech and hence the less twist.

- Different wind and sea conditions call for different amounts of twist. Light and very light winds require lots of twist. Medium and strong winds call for little or no twist. Heavy winds require you to reef the mainsail.

- Sailing in waves requires more twist and more power than sailing on flat water. Having more twist than you would on flat water helps keep the boat sailing as it is rocked around by the waves.

To prepare you for the rest of the book, Chapter 2 illustrates the mainsail, its place on the boat, the names of its controls and what they do.

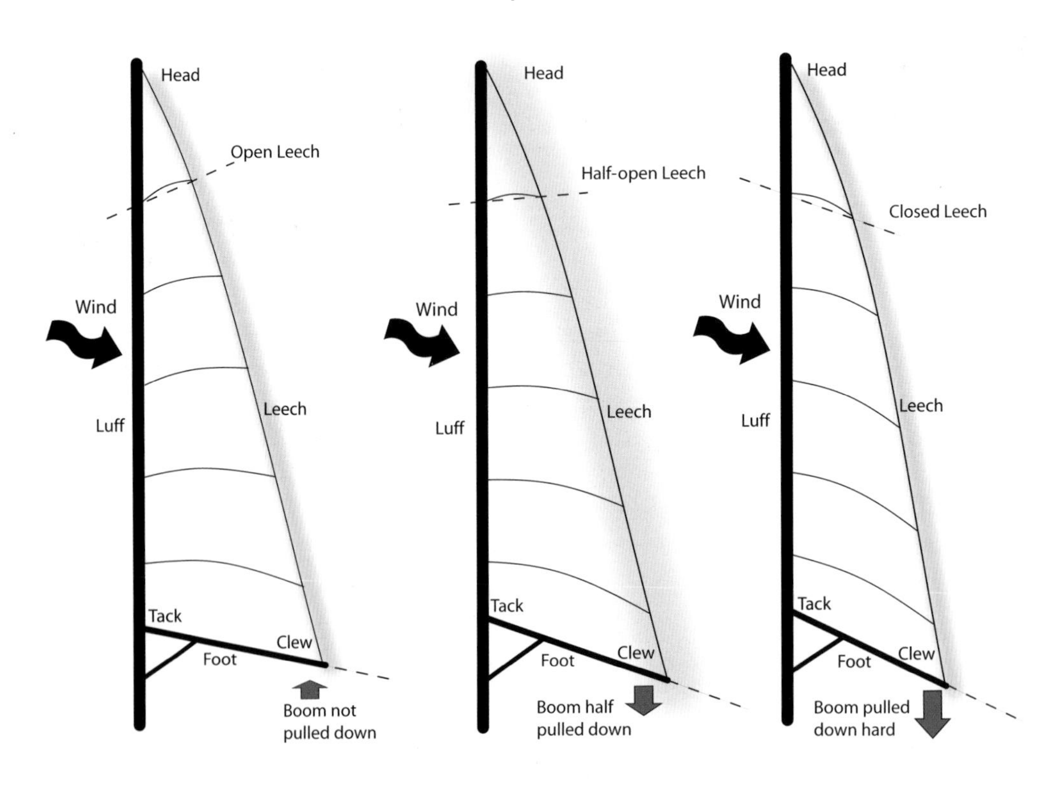

The mainsail and its controls

The Mainsheet

Traveler

Vang

Halyard

Cunningham

Outhaul

Backstay

In this chapter you will discover what each of the mainsail's primary controls is called, what it does, and how and when you operate it. Once you are familiar with the function of each control, you will easily be able to understand the following chapters, which explain the sail shapes you need to create for each wind condition and point of sail.

The smaller the boat, the less mainsail controls there are. A dinghy might be equipped with a mainsheet, halyard, outhaul, vang (kicker) and cunningham. Larger boats, including yachts, will have these controls plus a few more, including reefing lines, and backstay.

The Mainsheet

The most obvious and best-known mainsail control is the mainsheet. The mainsheet is the name for the rope that is used to haul the mainsail in, or to ease it out. On some boats, the mainsheet attaches to the end of the boom. On other boats, the mainsheet attaches to the middle of the boom or somewhere in between. Its actual position entirely depends on the design of the boat as follows:

- Smaller boats will have the rope running though some blocks (pulleys) before it gets into your hands. These blocks are configured to act as gears and enable you to pull the sail in even when it's powered-up by the wind. The main block will, or at least should, feature a ratchet that makes it easier to keep the sail trimmed-on when the mainsheet is not cleated.

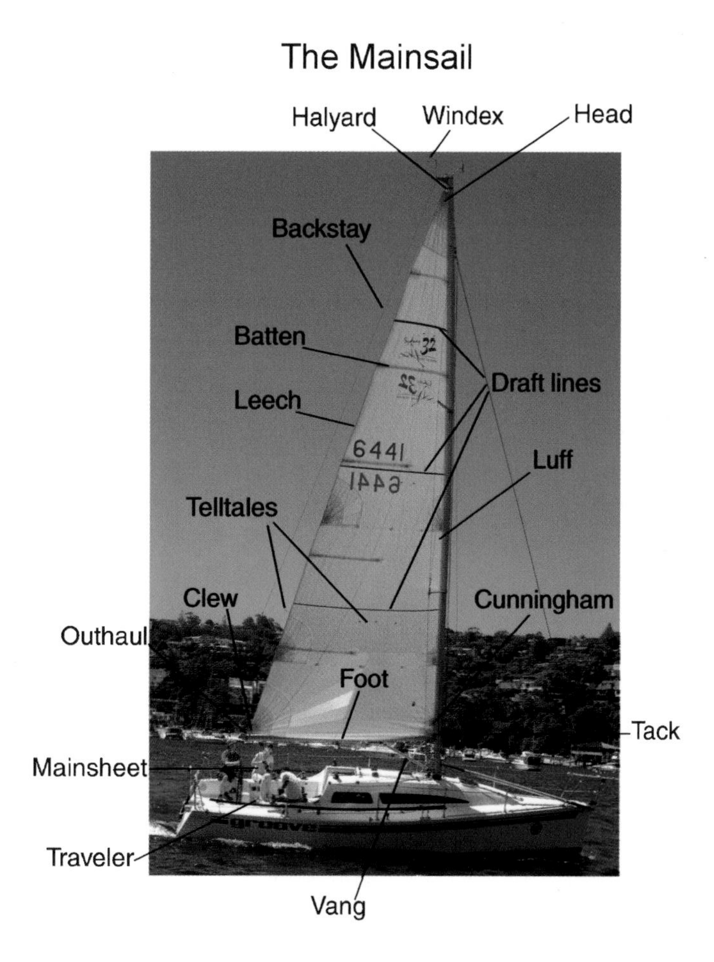

- Larger boats have more elaborate assemblies, all of which are designed to make it easy (and possible) to trim large sails that are fully powered-up.
- Trimming-on the mainsheet brings the boom, and hence the sail, in towards the boat's centreline.
- Easing the mainsheet lets the boom move out away from the centreline eventually to a position where the boom is perpendicular to the boat's centreline.

- Trimming the mainsheet on hard whilst sailing close-hauled encourages the boat to point higher (closer to the wind). However, trimming-on too hard whilst sailing slowly, prevents the boat from accelerating. Under normal conditions, it is necessary to accelerate before pointing high.
- Easing the mainsheet a little whilst sailing close-hauled makes it possible to accelerate.
- Easing the mainsheet quickly is necessary to allow the boat to bear away.

As you will shortly see, when the mainsail is close-hauled (trimmed on hard with the boom near the centreline and the traveler set low – see description of traveler in the following section), the mainsheet pulls down hard on the boom. This has the effect of closing the leech of the sail – as well as keeping the sail close-hauled. If the traveler is high, only a small amount of mainsheet tension is required to get the boom to the centreline, in which case trimming-on the mainsheet exerts little down force.

Understanding this general point is fundamental to being able to control twist and is the basis for how the traveler works.

Mainsheet Settings for Each Point of Sail

Let's take a look at the mainsheet setting for each point of sail (see Appendix C for an illustration of the points of sail).

Close-hauled

You sail close-hauled when you're sailing as close to the wind as possible. This is also referred to as beating or working.

- The boom's position is set close to (or on) the boat's centreline. The boom's ideal exact position will depend on the type of boat and the strength of the wind. For example, single sail dinghies will usually be set below the centreline. Boats with a mainsail and a jib will usually have the boom set higher up the track.

On boats without travelers, s, e.g. dinghies, the mainsheet is used for speed and heel control. This is discussed in detail later in the book and also mentioned under the description of the traveler in this section. The main concepts to understand at this stage though are as follows:

- Excessive heel resulting from increased wind (i.e. a gust or wind at the top of a wave) slows the boat down.
- If you're sailing close-hauled and you're hit by a gust, the boat will heel over more than it was heeling before.
- If the heeling is excessive, you can reduce the effect of the gust by quickly and slightly easing the mainsheet. As the gust passes, you trim the mainsheet on again.
- Maintaining a consistent angle of heel is a fast way to sail.

Reaching

While you're reaching, and since the mainsheet exerts less down force on the boom, you use the vang to control the leech shape (twist).

- When close reaching, the boom is close to the centreline.
- When beam reaching (wind coming towards the middle of the boat), the boom is set roughly halfway out.
- When broad reaching, the boom is set at roughly three quarters of the way out.

Running

When running, the boom is set all the way out so the boom is perpendicular to the centreline of the boat.

Traveler

Not all boats have a traveler. Small boats such as dinghies tend not to have travelers. Larger boats such as yachts usually do have travelers.

The traveler is used, whilst sailing close-hauled, for two important and distinct purposes. The first is to control the angle at which the mainsheet pulls the boom in (which in turn affects leech shape). The second is to make the boat go fast and stay on its feet.

Leech Shape

The traveler works with the mainsheet to control leech shape. Different wind strengths and sea states require different sail shapes. When to use these different sail shapes is explained later on. For the time being though, twist (leech shape) is the amount that the leech at the top of the sail is angled out from the centreline of the boat when compared with the boom. The greater the angle, the more twist there is. The smaller the angle, the less twist there is.

As mentioned earlier, and when you are sailing close-hauled, the mainsheet exerts a lot more down force on the boom than the other points of sail. The more down force there is, the more closed the leech will be and hence the less twist there is.

- Moving the traveler up towards you (whilst easing the mainsheet to keep the boom on the centreline) increases the angle that the mainsheet is pulling from, which reduces the down force on the boom – and therefore increases twist.

- Moving the traveler down away from you (whilst trimming-on the mainsheet to keep the boom on the centreline) sets the mainsheet more directly under the boom and hence causes the leech to close – and therefore reduces twist.

In other words, the traveler is fine-tuning the amount of down force that the mainsheet is exerting whilst you are sailing close-hauled and therefore controls how much twist there is in the sail.

Remember, the traveler is only really used to control twist in the sail whilst you're sailing close-hauled. As soon as you start reaching, you change over to the vang.

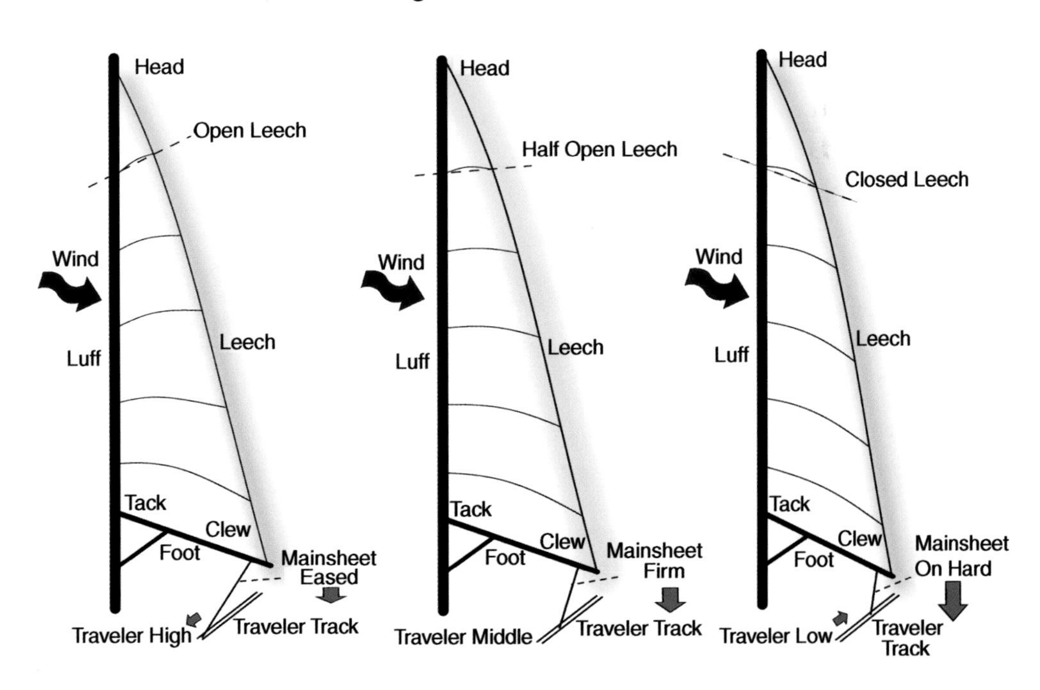

Speed and Heel Control – Very Light, Light and Medium Airs

The second function of the traveler is to provide heel and speed control. For boats without travelers, this function is performed by the mainsheet as described above.

This topic is discussed in greater detail later in the book. However, the main concepts to understand at this point are as follows:

Medium Twist - Traveler Middle

- Excessive heel resulting from increased wind (i.e. a gust or wind at the top of a wave) slows the boat down.
- If you're sailing close-hauled and you're hit by a gust, the boat will heel over more than it was heeling before.
- If the heeling is excessive, you can reduce the effect of the gust by quickly easing the traveler down the track. As the gust passes, you bring the traveler back up again.
- Maintaining a consistent angle of heel is a fast way to sail.

No Twist - Traveler Low

As you're hit by a gust, you could ease the sail out with the mainsheet (instead of the traveler). However, the advantage of using the traveler is that the down force on the boom remains about the same as you ease the traveler, which prevents the leech from opening. If you had eased the mainsheet instead, the down force on the boom would have lessened, the boom would have lifted and the leech would have opened. This would have slowed the boat down.

Speed and Heel Control – Strong and Heavy Airs

When the wind builds beyond a certain strength, the traveler is set down at the bottom of the track. Therefore, there's no more track to ease the traveler down during gusts. At this point, it becomes necessary to ease the mainsheet instead of the traveler.

Traveler

Traveler in hand
Mainsheet
Traveler car
Traveler track

In heavier airs, to prevent the boom from lifting and therefore the leech from opening as you ease the mainsheet, it's necessary to set the vang on – a technique called vang sheeting – see the description of the vang below.

Vang

The vang is used to control the leech shape (twist) by exerting only down force on the boom. When the vang is in use, the mainsheet's role is relegated to simply pulling the boom in or letting it out.

On dinghies, the vang is often called the kicker. On larger boats, the vang is also referred to as the boom vang.

The vang's use depends on the point of sail and the type of boat. There's a surprising amount to know about the vang so we've covered the vang's use for each point of sail with differences highlighted between boats with and without travelers.

Close-Hauled

As mentioned above, while you're sailing close-hauled, the traveler (not the vang) is used to control leech shape in anything but strong and heavy airs. As a general rule of thumb, when sailing in light to medium airs, always ease the vang as you change course upwind.

Boats with Travelers, e.g. Yachts

As we also said above, one of the traveler's two primary functions is to keep the boat going fast by easing in gusts. However, in heavier airs you cannot ease the traveler in a gust because the traveler is already at the bottom of its track. Therefore, the vang is set and the mainsheet becomes the only control available to respond to gusts – this is called vang sheeting, i.e. in gusts, you keep the vang on and ease with the mainsheet, not the traveler.

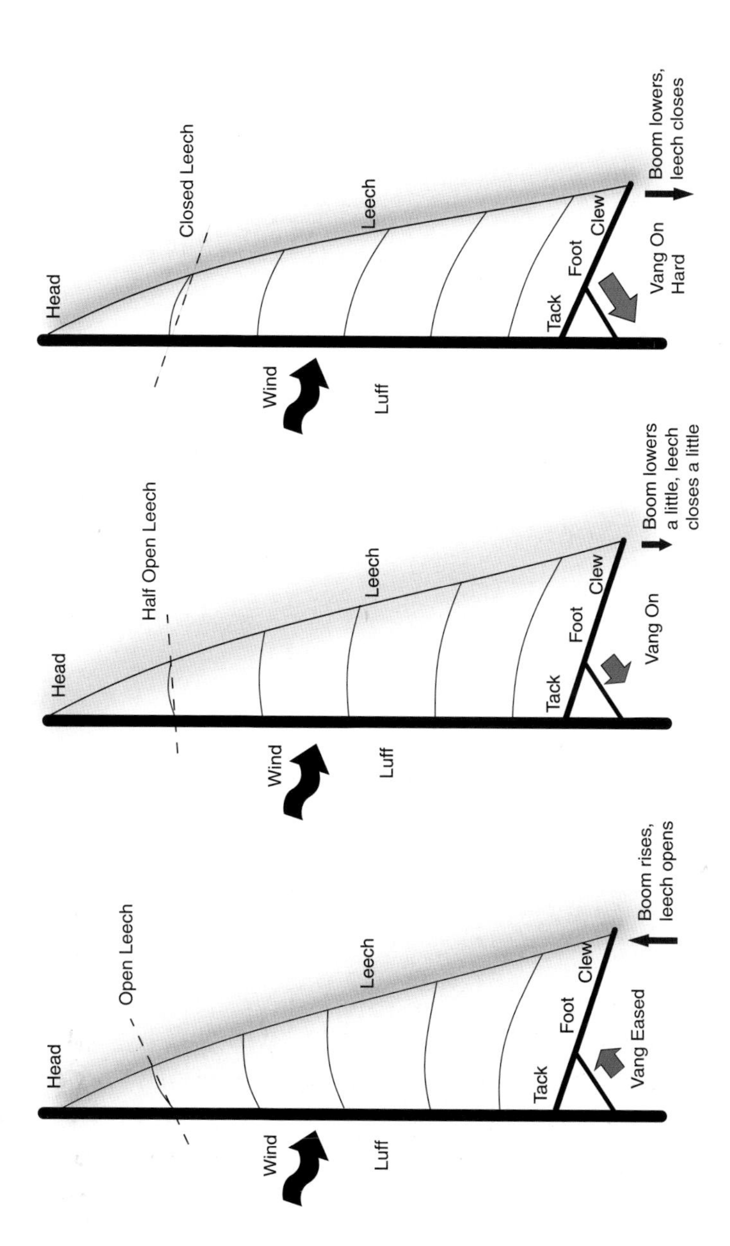
Controlling Twist with the Vang
Head
Open Leech
Leech
Clew
Foot
Tack
Vang Eased
Boom rises, leech opens
Wind
Luff
Head
Half Open Leech
Leech
Clew
Foot
Tack
Vang On
Boom lowers a little, leech closes a little
Wind
Luff
Head
Closed Leech
Leech
Clew
Foot
Tack
Vang On Hard
Boom lowers, leech closes
Wind
Luff

Boats without Travelers, e.g. Dinghies

On boats without travelers, or with travelers that cannot be moved easily, the vang is used for all points of sail, including close-hauled. In this case, vang sheeting is the norm.

Sailing upwind without a traveler still requires the same control over the leech as we saw in the last section on the traveler – but you use the vang instead. In light winds you need lots of twist (light vang tension), in medium winds firmer tension, and in strong winds you need very firm tension to completely close the leech.

Please examine the photos and illustrations in the previous section to see what leech shapes you need to develop with the vang for upwind sailing. The following illustrations and pictures show how to use the vang.

The absence of vang tension allows the boom to rise and the sail to fall open

The vang pulls the boom down and closes the leech

Reaching and Running

When reaching or running, the vang is always used on all types of boat. Since the vang holds the boom down (controlling the leech shape, i.e. the twist), the mainsheet simply pulls the boom in or lets it out.

Close Reaching and Beam Reaching

The approximate setting for the vang is the tension required to set the top batten parallel with the boom. The exact setting will depend on the strength of the wind.

- When close reaching and beam reaching in light and very light airs, the sail will need twist and so the vang should be set with relatively little tension. This causes the top batten to be angled out from the boom by a few degrees. Use the telltales to determine whether the sail is over-sheeted or under-sheeted. See the section on telltale reading in Chapter 3 for an explanation of how to read the telltales.
- In medium and stronger airs the vang should be pulled on hard to eliminate twist.
- When reaching in stronger airs, easing the mainsheet during gusts may not be enough to stop the boat from broaching. When you're hit by a gust and you're close to broaching, your only defense will be to ease the vang. In these conditions, the vang control should always be in someone's hand, preferably while they are still hiking on the rail, so the vang can be eased at a moment's notice.
- There may also be times when simply easing the mainsheet is not enough to avoid a collision. In an emergency dip or bear-away, the vang might be the last hope for the helmsman to be able to steer safely.

Broad Reaching and Running

While broad reaching and running, the vang should be set to keep the top batten roughly parallel to the boom, i.e. there should be little or no twist.

On some boats in light airs, there will be too much weight in the sail and boom, and hence the leech will be too closed – even without the vang being set. If this is the case, consider having a solid vang fitted. Solid vangs are sprung and resist the boom from being pulled down under its own weight.

Halyard

The halyard is the rope attached to the head (top) of the sail. It is used to hoist the sail.

The halyard and the cunningham both control the tension in the luff of the sail. Both controls effect the completeness of the sail's shape but in particular, the position of the belly of the sail. The halyard and cunningham also flatten the sail if they're pulled on tight.

- More halyard tension causes the sail's belly to move forward towards the mast and the sail to become flatter.
- Less halyard tension causes the belly to move backwards – away from the mast. The goal with the mainsail is to get the belly in the middle of the sail.
- Using too little halyard tension means the sail will not be fully hoisted.

See Chapter 6 on hoisting, dropping and reefing for further information on how to hoist and drop the mainsail.

Cunningham

On boats with backstays, the cunningham is attached to the sail's luff a little way above the boom and is primarily used to take out the sag from the sail when backstay tension is applied.

- As the backstay tension is increased, the belly moves back (away from the mast) and the sail appears as if the halyard has slipped down a little. This is because the mast is bent by the backstay, which reduces the distance between the top of the sail (head) and the tack (the sail's bottom corner

Cunningham

Cunningham off

Tensioning backstay without tightening the cunningham causes the luff to sag

Cunningham on

at the mast). In other words, the sail sags and the sail's overall shape deforms.

- The cunningham is used to take up this slack so that the sail regains its shape.

On dinghies, where there is typically no backstay, pulling on the cunningham is one of the primary ways to flatten and de-power the whole sail by flattening the bottom of the sail and completely opening the leech.

On some yachts, the cunningham is often used as a tack line when the sail is reefed – again, see Chapter 6 on hoisting, dropping and reefing.

Outhaul

The outhaul is used to control the size of the sail's belly in its lower third.

- The tighter the outhaul, the flatter the sail's shape in the lower third.
- Easing the outhaul increases the size of the belly and hence increases its power. The outhaul is never eased more than around 15% of the overall length of the sail's foot.

The size of the belly in the upper two thirds of the sail is controlled using the backstay – see later in this chapter for details on how to use the backstay.

- In very light airs, the outhaul should be fairly tight to create a flat foot. This is to allow the weak wind to attach to the sail. If the foot is too big, the wind won't make it around the sail at all.

- In light airs, the outhaul should be eased to the maximum setting (around 15 % of the length of the foot). This creates the most powerful shape.
- In medium airs, the outhaul should be brought on a little to start to flatten the foot of the sail.
- In stronger airs, the outhaul should be brought on hard to keep the foot flat. This is achieved by tightening the outhaul until the clew meets the black band – see the following photograph.
- Tightening the outhaul reduces the belly in the lower third and so reduces power.
- Easing the outhaul increases the belly in the lower third of the sail and so increases power.

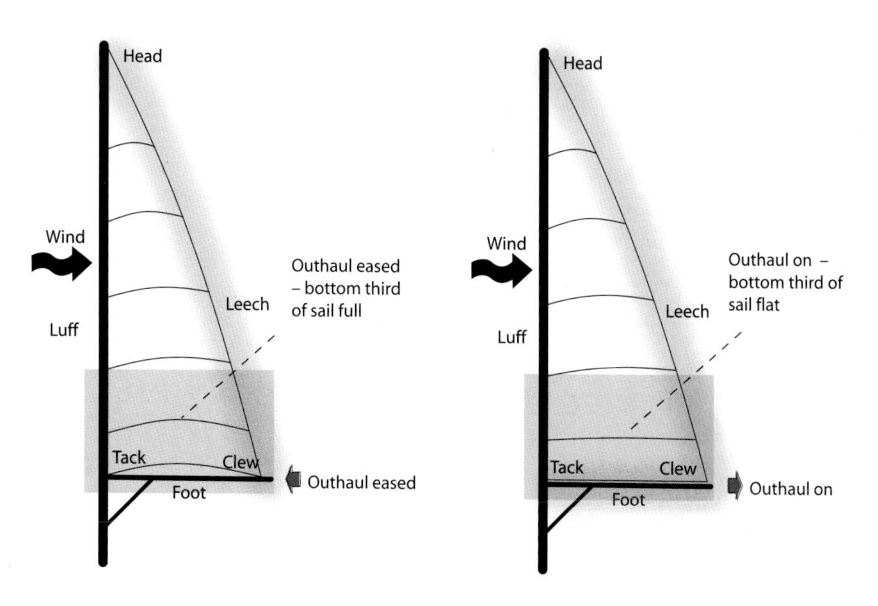

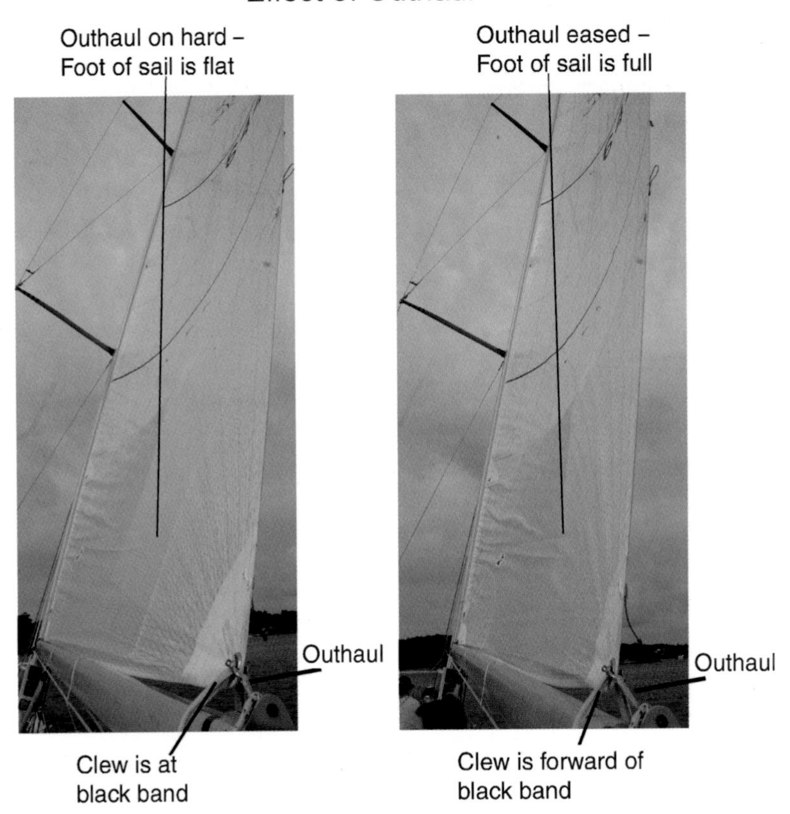

Backstay

The backstay is used to bend the mast and, in so doing, flatten the belly in the top two thirds of the sail. This is done to reduce the power in the sail. As we said above, the outhaul is used to control the belly size of the sail in the lower third.

- With no backstay tension, the top two thirds of the sail is at its fullest and most powerful (within the constraints set by rig tension).

- With full backstay tension, the top two thirds of the sail is at its flattest and least powerful (again, within the constraints set by rig tension).
- With excessive backstay tension, diagonal lines appear in the sail that run between the clew and the luff.

The backstay can be adjusted throughout a race and is essential for keeping the boat under control.
There are many different backstay arrangements, including block and tackle and hydraulic mechanisms:

- The backstay is only useful whilst the sail is acting as an airfoil, i.e. close-hauled and reaching.
- The backstay is used in very light airs to flatten the top two thirds of the sail so the wind can attach to the sail without stalling. At the same time as the backstay is tensioned, trim the outhaul on to create the same effect in the lower third of the sail.
- Although the backstay is used for very light airs, it is not used for light to medium airs. You start to use the backstay again in medium and heavier airs. The windier it is, the more backstay you use. It's also critical for speed to remember to ease the backstay once the wind drops in strength
- Always consider easing the backstay when you alter your course away from the wind to a broad reach.

Understanding exactly how much backstay to use is covered in Chapters 3 and 4. As a general rule though, don't start flattening the sails (outhaul and backstay) until you're over powered and you have everyone hiking out.

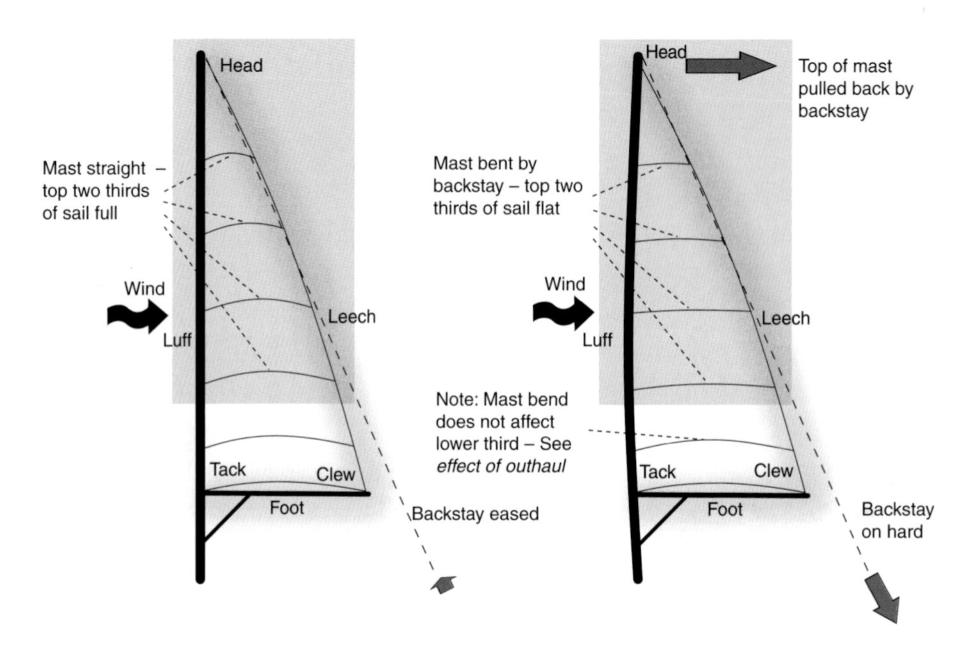

The backstay limits overall power (within the boundaries set by the rig tension) but can be adjusted whilst racing.

- The harder the backstay is tensioned, the more the mast bends. Bending the mast flattens the top two thirds of the mainsail sail, which causes the top two thirds of the mainsail to become less powerful.
- The looser the backstay, the straighter the mast, the fuller the sail in the top two thirds, and so the more powerful the top two thirds of the sail becomes.

As indicated in the first section of this book, an important point to note is that the backstay and outhaul can only operate within the overall context of rig tension. The following illustration shows the combined effect of backstay and outhaul.

Effect of Backstay, Outhaul and Rig

Backstay and outhaul eased Loose rig

Draught lines rounded, big depth = powerful sail

Backstay and outhaul on Loose rig

Draught lines less rounded, less depth = less powerful sail

Backstay and outhaul on Tight rig

Draught lines hardly rounded, little depth = least powerful sail

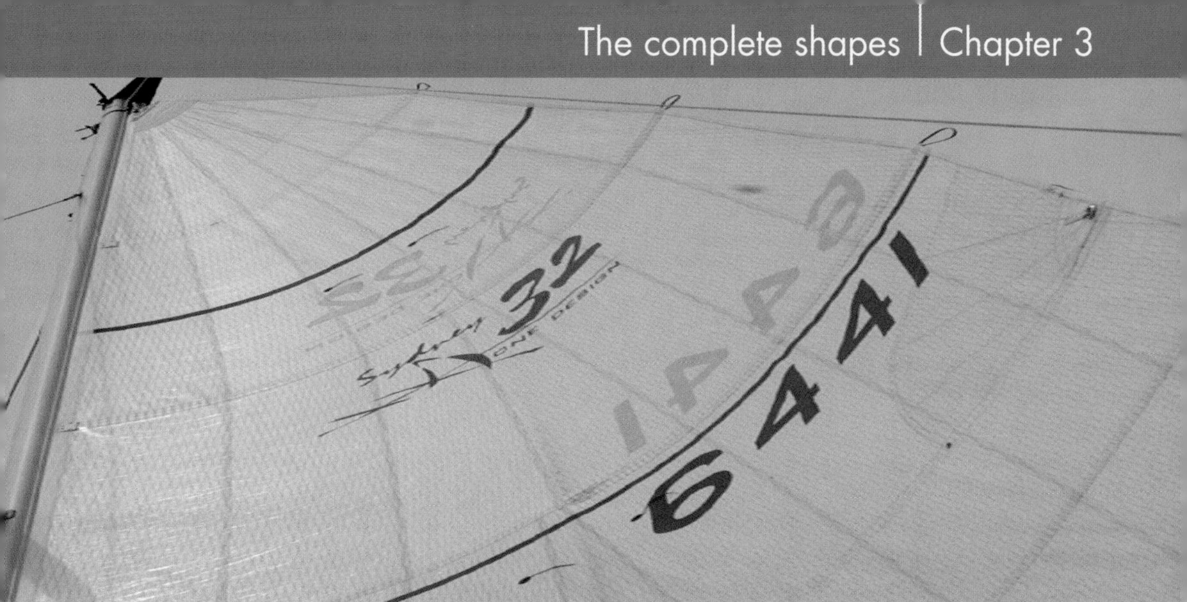

The complete shapes

The Complete Upwind Shapes

Reading the Telltales

Now that you have understood the concepts of belly size and twist, and the mainsail controls and when to use them, it's time to put them together.

The Complete Upwind Shapes

For very light airs to storm force winds:

- In very light airs, the wind will be struggling to get around the sail. You create a flattish shape (to help the air attach to the sail) with lots of twist. The outhaul should be trimmed half on to create a flat foot. The backstay should also be tightened on to make the sail flat in the upper two thirds. The leech must be twisted so the traveler should be at its highest setting and the sheet should be eased.
- In light airs, the wind will be able to get around the sail, so maximum belly and twist are required. The backstay should be completely off and the outhaul eased to its maximum position. The leech must be twisted, so the sheet should be eased and the traveler should be high.
- As the wind builds to medium airs, you no longer want full power or twist in the sail. The backstay and the outhaul both need to be brought on as soon as everyone's hiking and you're still over powered. The amount of outhaul tension and backstay tension depends on the strength of the

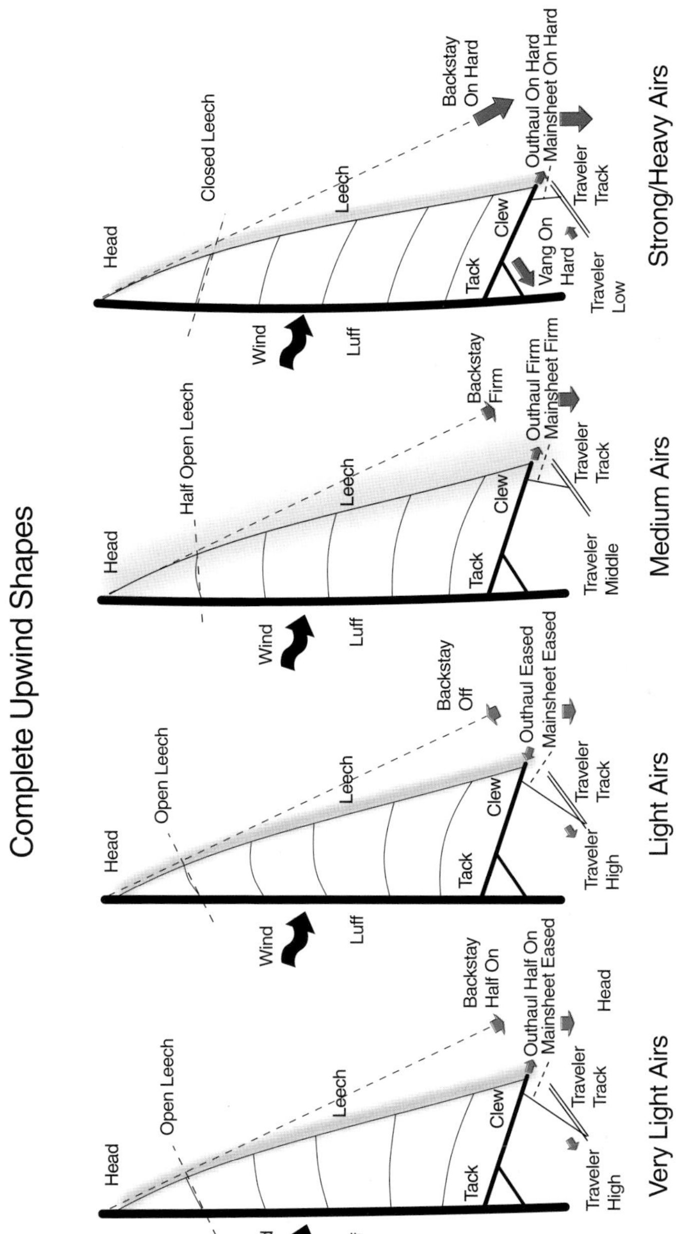

Complete Upwind Shapes

Very Light Airs

Head
Open Leech
Leech
Clew
Tack
Luff
Wind
Backstay Half On
Outhaul Half On
Mainsheet Eased
Head
Traveler Track
Traveler High

Light Airs

Head
Open Leech
Leech
Clew
Tack
Luff
Wind
Backstay Off
Outhaul Eased
Mainsheet Eased
Traveler Track
Traveler High

Medium Airs

Head
Half Open Leech
Leech
Clew
Tack
Luff
Wind
Backstay Firm
Outhaul Firm
Mainsheet Firm
Traveler Track
Traveler Middle

Strong/Heavy Airs

Head
Closed Leech
Leech
Clew
Tack
Luff
Wind
Backstay On Hard
Outhaul On Hard
Mainsheet On Hard
Traveler Track
Vang On Hard
Traveler Low

wind. To start with (as the wind starts to build in strength) only a little of both is required. As medium airs become stronger, you will need to bring both on more and more. How much of both will become clear in the next section and is one of the most important things to understand about mainsail trimming. The leech should be fairly closed so the sheet should be trimmed on firmly and the traveler should be somewhere in the middle of its range.

- In strong winds, a flat sail with no twist is needed, so the backstay and outhaul should be on hard. The traveler should be down the bottom of the track and the vang should be on hard to keep the leech closed as the mainsheet is eased during gusts. If the boat is still over powered with the sail completely flat, then it's time to reef.
- In gale and storm force winds, you need to rig a tri-sail and a storm jib! Apart from being practiced and equipped, the most important thing to know about the tri-sail is that it must be rigged and set so that the belly is flat. Setting a powerful storm sail is completely self-defeating!

Reading the Telltales

Telltales are indispensable for providing a visual indication of how well the sail is trimmed whilst you're sailing close-hauled and reaching, i.e. when the sail is acting as an airfoil rather than an air dam.

Telltales can immediately tell you whether you're over, under or correctly trimmed, and whether the sail's twist is correct. Note that they don't tell you whether the sail is under or over powered.

There are two types of telltale that can be attached to a mainsail: leech telltales and draft telltales. It's preferable to have both types attached.

Leech Telltales

As their name suggests, leech telltales are attached to the sail's leech. Leech telltales work by showing whether the sail is over trimmed or under trimmed as follows:

- If the sail is over trimmed, the telltales will wrap around the outside of the sail.
- If the sail is under trimmed, the telltales will stall or wrap around the inside of the sail.
- If the sail is trimmed correctly, the telltales will flow out evenly.

Reading the Telltales

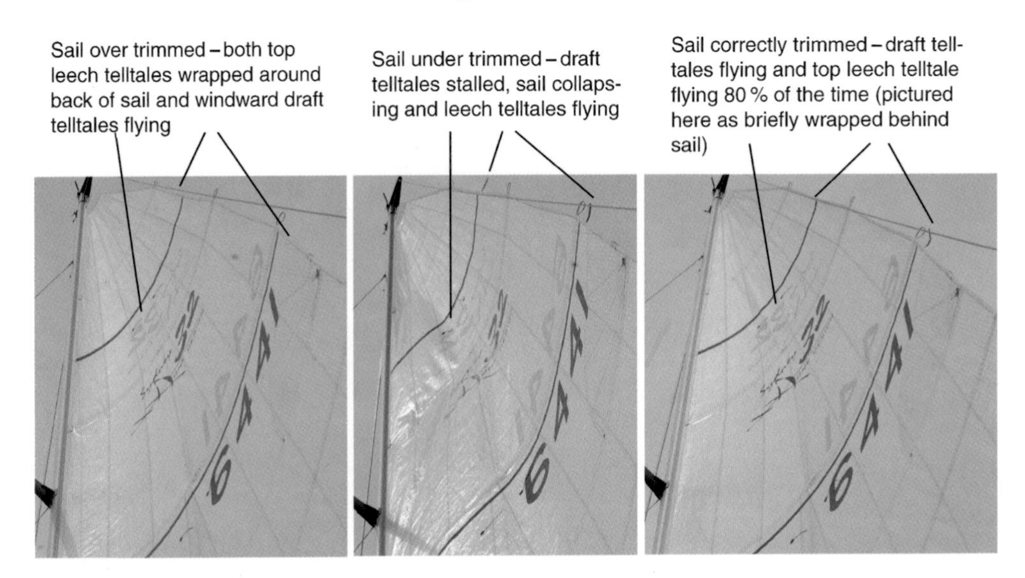

The telltales also provide an indication of whether the twist is correct. Assuming that the bottom and middle leech telltales are flying:

- If there's not enough twist, the top telltales will wrap around the outside of the sail.
- If there's too much twist, the top telltales will stall or wrap around the inside of the sail.
- If the twist in the sail is correct, the top telltales will flow out evenly for roughly 80% of the time.

Draft Telltales

Draft telltales provide a better level of visibility into where exactly the air is flowing on the sail.

- If the sail is over trimmed, the outside telltales will lift and stall whilst the inside telltales fly.
- If the sail is under trimmed, the inside telltales will lift or stall whilst the outside telltales fly.
- If the sail is trimmed correctly, both the inside and outside the telltales will fly.

As we saw above for leech telltales, the draft telltales also provide an indication of whether the twist is correct. Assuming that the bottom and middle inside and outside draft telltales are flying:

- If there's not enough twist, the top outside telltales will lift and stall whilst the inside telltales fly.
- If there's too much twist, the top inside telltales will lift or stall whilst the outside telltales fly.

If the twist in the sail is correct, both the top inside and outside the telltales will fly.

How to sail fast upwind

Gears

Pointing only Comes with Speed!

Keeping Speed

Controlling Heel

Using 'Feel'

Using Target Boat Speeds

Fast Maneuvers

So far in this book, we have shown how lift and drag are created and managed. From this foundation, we have carefully examined each of the mainsail controls available to optimize sail shape, and hence lift and drag.

Using an automobile analogy, we've covered the concept of cars, roads and controls such as brakes, accelerators and windscreen wipers. We haven't yet explored how to drive in the real world with bendy roads, steep hills and loose surfaces. This section of the book is dedicated to the actual process of driving the boat using your control over the mainsail.

The difference between a competent and a brilliant mainsail trimmer is enormous. Making the boat accelerate fast, keeping top speeds and maneuvering around a course requires complete concentration. Not only do you need to be able to make any change to any aspect of the sail, you need to anticipate the need to make the change. If you think it is okay just to get the right sail shapes whilst you traveling at top speed and do nothing more, you're off the money by at least 50% of the boat's potential.

There are several major concepts and techniques to understand before you can sail fast. These include straight-line sailing techniques such as using gears, heel, feel, and target boat speeds. These concepts also include how to execute fast maneuvers for heading up, bearing away, tacking, footing, responding to wind changes and sea state.

This is a long chapter but it contains all the key concepts you need to make the boat go fast. All of these techniques will be employed by the fastest boats on a racecourse. Collectively, they're the go-fast techniques that the pros use all the time to keep at the front of the fleet.

We recommend that you keep coming back to this section until you're sailing fast all of the time.

First though we'll start, using the car analogy, by explaining how to use gears.

Gears

It's extremely useful to think of sailing as you think of using gears in a car. If you try to accelerate from 10 mph in top gear, you will fail, or at least take a very long time to go faster. It's the same with sailing. Top speed trim is different from starting trim. You can only get to top speed by going through two lower gears: starting and accelerating. Therefore, there are three main gears, and there's an overdrive too if you're sailing on flat water in a good breeze.

In order to be able to change gears, you need to know which gear you're in at any given time. This will help you get the trim right immediately. As a rule of thumb, if you're in any doubt, the chances are that you're over trimmed.

Remember that to achieve top speed, you must go through each gear – you can't jump from 1st to 3rd gear. Trying to do so (by being trimmed for 3rd gear whilst you're traveling at a 1st-gear speed) means that you won't accelerate. What might fool you into thinking you can go from 1st to 3rd is that occasionally, gusts can help you to accelerate whilst you're over trimmed. However, you can't wait around for gusts, you need to accelerate immediately!

1st Gear: when starting from scratch; at race start; after a large and unexpected wave and, after a bad tack:

- You ease the sheet a couple of inches and the helmsman steers down (away from the wind) until the middle telltales are flowing on the inside and outside of the jib.
- You stay in this gear until you've built up some speed, perhaps for 15 seconds.

2nd Gear: after a good tack; after some waves that you've successfully footed through; at the start; during a lull and, after 1st gear):

- You trim the sheet on a bit more and the helmsman steers up (closer to the wind).
- You stay in this gear until you've built up some more speed, perhaps another 15 seconds.

3rd Gear: after you've accelerated from 2nd gear; after a very good tack, and at the start):

- You trim the sail on fully and the helmsman steers up (towards the wind) until the outside telltales on the jib are flowing evenly and the inside telltales are flowing upwards at around 45 degrees.

Overdrive Gear: (after you've made it into 3rd gear).
This is our overdrive gear and is referred to as 'sailing lifted':

- On flat water and in good wind, the helmsman can steer even higher (closer to the wind), so the inside telltales are pointing almost vertically. The mainsheet can be even further tightened. However, as soon as there's a lull or some chop, you will immediately need to change down a gear to keep the boat speed up by easing out slightly.
- Whether you can reach 2nd, 3rd and overdrive gear depends on the wind strength and sea state. If the wind is very light, you will stay in first gear. If the wind is just light, you might make it to 2nd gear. Only if there's enough wind, e.g. over 10 knots, will you be able to get into 3rd gear. Overdrive is only achievable in winds over approximately 14 knots and on flat water.

Measuring Gears

Unsurprisingly, you are not the only trimmer on a two-sail boat. The jib trimmer needs to be using these gears at the same time as you. To facilitate using the gears properly, it pays to be vocal about which gear you think you're in and whether or not you think it's time to change up or down.

A useful technique is for either you or the jib trimmer to be responsible for calling the gears. On a boat with a boat speed indicator, the whole process can be managed as follows:

You establish which boat speed ranges equate to each gear. Here's an example, though the real figures vary for each type of boat:

- 1st gear: 0–2 knots boat speed;
- 2nd gear: 2–4 knots boat speed;
- 3rd gear: 4–6 knots boat speed;
- Overdrive: 6–6.7 knots boat speed.

Now, as you accelerate through these gears, you or the jib trimmer can be responsible for calling the gears based on actual boat speed. This approach ensures the following:

- You coordinate with the jib trimmer;
- You use gears;
- You don't try to jump gears;
- You accelerate fast.

Pointing only Comes with Speed!

Pointing is a measure of how close you are sailing to the wind. Boats typically sail at around 37 degrees off the wind when they're in 3rd or overdrive gear, though the exact figure varies depending on the design of the boat.

The important thing for you to understand is that you can't point as high as you can in 3rd gear whilst you're still in 2nd or 1st. Pointing only comes with speed. Therefore, you must expect to point badly until you get up to speed. If you're pointing well and going slowly, you are sailing too high, your sail is over trimmed and you are not accelerating. Understanding this concept alone will ensure you go faster than many other boats on the water.

One of the simplest generally available tools for assessing how well the boat is pointing is the wind indicator at the top of the mast. This instrument is used by the helmsman, tactician, trimmers and anyone else with an interest in the current direction of the apparent wind.

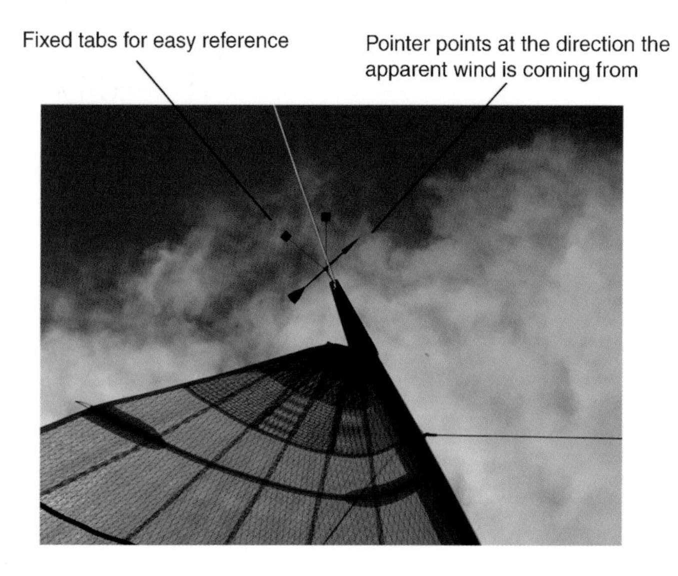

In this picture, the wind is coming from the side of the boat, i.e. the boat is on a beam reach. The two fixed tabs are generally set at around 40–45 degrees from the centreline of the boat.

If the tabs are set correctly, it's possible to tell immediately how high you're pointing:

- If the tail of the wind indicator is in line with the tab, you should be in 1st, 2nd or 3rd gear.

- If the tail is inside the tabs, you're pointing high. Remember, this is only correct if you're sailing in overdrive gear (sail lifted), or sailing deliberately high for tactical reasons.
- If the tail is outside the tabs, you're sailing too far off the wind – unless of course you're deliberately footing, dipping, etc.

This picture shows the mainsail set close-hauled on starboard tack and the pointer is in line with the tab.

Keeping Speed

You need to keep speed to build on it. If your response to a gust slows you down, you will always be speeding up again trying to re-capture speed. Your goal is to keep speed and build on it until you get to the target boat speed. Once you're at the target, you must stay there.

Keeping this up all the time requires serious concentration and physical effort!

It's also possible to appear to achieve your target speed, or even exceed it by sailing too far off the wind. This is a better quality problem to have than not being able to achieve your target speed in the first place – but it's still bad – you're not trimmed on enough! When you're sailing close-hauled, your goal is not just to get to your target speed but to be pointing high enough when you do.

If your target speed is 6.7 knots and you achieve 7 or 7.2 knots, it's not because you're a genius and you have outsmarted the targets. It's because you're sailing several degrees off the wind, that is, further from your upwind destination.

The reason for going offwind will probably be because you haven't sheeted-in enough because you're over powered if you do. In other words, the sail is too powerful for the conditions. Therefore, you must de-power the sails.

As we've already seen in this book, de-powering is achieved by tightening the outhaul and backstay. If you're already at the maximum for both, you must consider putting in a reef.

Failing to identify that you're over powered will spoil your upwind progress by making the boat point badly. Failing to identify that you're under powered will make you slower and lose ground too.

Reacting straight away to changing conditions is the key to acquiring and keeping speed.

Controlling Heel

Heel is the degree to which the boat is tilted sideways. Maintaining the correct amount of heel is one of the most important aspects of upwind sailing. The importance of achieving the correct angle of heel cannot be over stated. It is of vital importance to sail at the correct angle of heel all of the time!

First, you must find out what your boat's ideal angle of heel is. Dinghies tend to need to be kept flat. Yachts generally need a reasonable angle of heel, e.g. between 15–25 degrees.

Heel should be constantly occupying your mind (and the mind of the helmsman) whilst you're sailing:

- As we saw earlier, you control heel by trimming-on and easing-off the traveler (or mainsheet if you're vang sheeting).
- If the ideal angle of heel is exceeded, you must ease the traveler or mainsheet (if you're vang sheeting).
- If the angle of heel is too low, you must bring up the traveler or trim on the mainsheet (if you're vang sheeting).

The consequence of sailing with the wrong angle of heel is extremely serious. The boat will sail much slower than its target speed for any condition. Also, if you sail at the correct angle of heel for only some of time, you will generally be going much too slowly.

Of all the many and very expensive gadgets you can buy for sailing, one of the simplest, cheapest and most useful is the inclinometer. This item usually consists of a plastic housing with a ball bearing running along a calibrated scale.

By using an inclinometer, you can easily and accurately see the angle of heel you're sailing at. Once you know your ideal angle, you can immediately judge whether you're sailing it. However, you should not spend all of your time looking at the meter. You should use it to learn what the correct angle looks and feels like. Once you have learned it, you should be thinking about whether your angle is right all of the time but only occasionally looking at the inclinometer to check.

Coordination with the Helmsman

Although this book only concerns itself with the mainsail, maintaining the right amount of heel requires absolute coordination with the helmsman.

There is absolutely no substitute for good communication between you and the helmsman. To make it possible for you to trim well, you need to know whether the helmsman is over powered. As soon as there's too much pressure in the helm, you must ease out. As soon as the excessive pressure starts to drop, you must start trimming-on again. You should be obsessing about heel to achieve this (see above).

A good indication of bad coordination is where the boat sails in zigzags with the occasional round-up. The boat should sail fast and straight!

The helmsman needs to tell you whenever there's excessive force on the helm. If

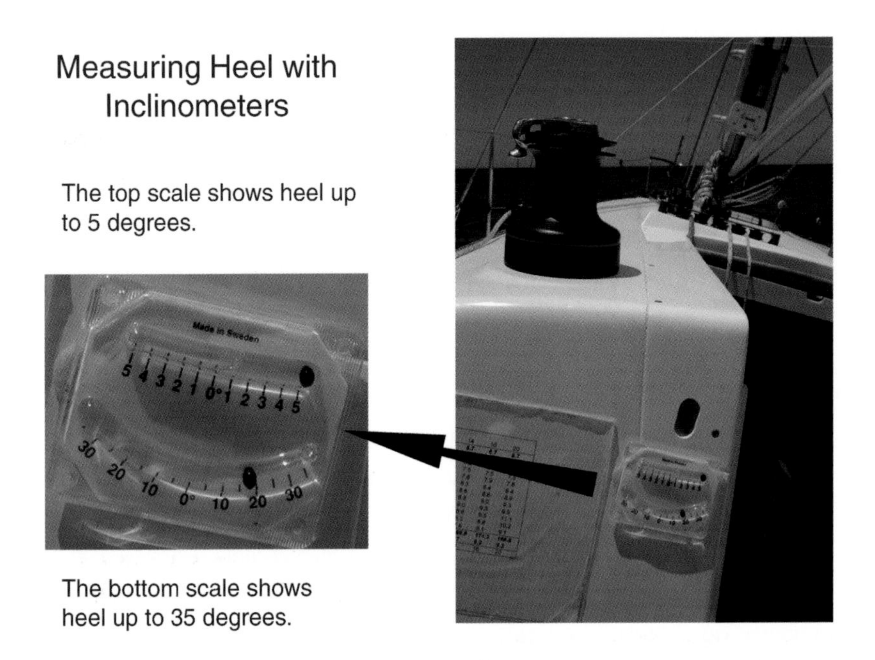

Measuring Heel with Inclinometers

The top scale shows heel up to 5 degrees.

The bottom scale shows heel up to 35 degrees.

the rudder is turned more than approximately 15 degrees to be able to steer straight, there's too much pressure on the helm and the boat will be slow.

To help matters, there are two good ready reckoners available (depending on the type of steering) that can tell you how much force there is on the helm. On a boat with a tiller, the angle of the tiller should correspond to the angle of the rudder. If the tiller's angle from the centreline of the boat exceeds around 15 degrees, you know the helmsman is over powered, even if you're not being told!

A similar method exists for boats equipped with steering wheels. A piece of visible tape is stuck to the top of the wheel when the rudder is completely straight. When you're sailing, the greater the force in the helm, the more the wheel is turned and the further round the tape will go. A quarter turn of the wheel usually corresponds to 15 degrees of rudder angle, though it requires a little calibration to be sure.

Very Light and Light Airs

In very light and light winds, the heel of the boat is created by the crew being located on the low side of the boat. As the wind strength builds, more crew weight is brought up onto the high side of the boat.

Medium and Stronger Airs

As soon as there's enough wind (medium airs and above), the helmsman and you work to create and maintain heel.

If the helmsman is pointing too high (too close to the wind) or too low (too far off the wind) the boat will be going slowly. If you trim-on the main to try to make the boat heel more when the boat is being steered too high, the boat will go even slower. Therefore, it's essential for the helmsman to be sailing correctly before you're able to achieve and maintain the right degree of heel.

For your sake, we hope that the helmsman is sailing well enough to create sufficient opportunity for you to trim properly. However, if the helmsman is not sailing well, your efforts at achieving the right angle of heel will fail. A good indication of this occurring is that the boat will be sailing slow, that is, in 1st or 2nd gear but the inside telltales on the jib will be lifting. This shows that the helmsman is steering too high for the current gear.

Another indication of poor steering whilst sailing in 1st or 2nd gear is that the back part of the Windex on the top of the mast will be inside or well outside the tabs.

There are only two reasons why the helmsman should be thinking about pointing very high: the boat has already achieved 3rd gear and the helmsman is sailing lifted (overdrive gear), or the helmsman is trying to point higher for some tactical reason.

Using 'Feel'

There are degrees of perfection in mainsail trimming. Your goal is to trim perfectly all of the time.

Getting the right angle of heel only some of the time, not responding to gusts fast enough, not changing the backstay and outhaul for building or dropping breeze, not getting the traveler to the right setting immediately are all examples of imperfect trimming. The best sailors use their senses to anticipate changing conditions and are ready to make whatever trim changes are required in good time.

Most of your senses collect the information you need. The challenge is getting your brain to register the information and to act on it.

Apart from seeing changing conditions, there are other very useful ways of getting some last minute warning about changing conditions. First, you can feel gusts on your face. Next time you go outside, stand still and wait for a gust to blow over you. Just think about the feeling the wind creates on your cheeks. Once you're attuned to this feeling, make it one of the things to think about next time you go sailing.

Another useful sense is that of balance. When you sit on a boat whilst you're sailing, you naturally balance yourself so you're comfortable. As soon as the boat starts to heel (more or less), you are no longer comfortable because you're using more effort to fight against gravity to keep you upright. Your body knows this straight away but your challenge is to use this information immediately for trimming purposes.

Simply waiting until the boat is heeled over 15 degrees more than the optimal level is not good enough. You need to anticipate and respond as much as you can.

If you can keep your head out of the boat and see what's coming, you can be ready for it. Here's an example:

Gust Approaching (what not to do)

You're sitting looking at where you've just come from. A gust hits without warning. You get the traveler down (or the sheet off) after the boat has heeled right over and screwed the bow up into the wind. The boat slows down, the helmsman steers back onto course and the boat builds up speed again over the next two minutes, just in time for the next big gust!

Gust Approaching (what you should do)

Here's what to do: you can see a gust approaching, you are ready to ease the traveler down (or ease the mainsheet) exactly as the gust hits the boat. In fact you can be ready to capture the power of the gust as it hits whilst easing to keep the heel through the gust, and then be trimming back on as the gust fades. Your boat speed stays at its target before, during and after the gust.

Of course, on a boat sailed by more than one person, it doesn't have to be you spotting the gust. It's always best if someone else on the boat is nominated to call gusts so everyone knows what's about to happen – and is ready for it!

Using Target Boat Speeds

For upwind sailing, target boat speeds are primarily for you, the mainsail trimmer.

Target boat speeds are theoretical speeds through the water that a boat can achieve for a range of true wind speeds and true wind directions. If the jib is set correctly and the helmsman is steering in a straight line with the jib telltales flying appropriately, it's you that gets the boat to its target speed.

Here's a sample table. The first row refers to true wind strength. The second, third and fourth rows assume you're sailing close-hauled. The second row indicates your boat speed through the water and the third row your apparent wind angle (pointing) you should achieve for each true wind speed.

The other rows are for offwind points of sail from 52 degrees upwards. It's a great idea for you to get a chart for the boat you sail on and paste it into the cockpit.

Example of Target Boat Speed Table

True Wind	6	8	10	12	14	16	20
Upwind Vs Upwind Bt	5.3 43.6	6.2 42.8	6.5 40.7	6.7 39.4	6.7 38.8	6.7 38.6	6.7 39.2
Upwind VMG	3.8	4.6	5.0	5.1	5.2	5.3	5.2
52	5.9	6.9	7.3	7.4	7.5	7.5	7.5
60	6.3	7.2	7.6	7.7	7.8	7.9	7.8
75	6.7	7.5	7.9	8.2	8.3	8.4	8.4
90	7.2	7.8	8.0	8.2	8.5	8.8	8.9
110	7.3	7.9	8.3	8.6	8.8	9.0	9.3
120	7.0	7.8	8.3	8.7	9.0	9.3	9.9
135	6.1	7.2	7.9	8.4	8.9	9.5	11.1
150	5.0	6.2	7.1	7.7	8.3	8.8	10.2
Run VMG	4.3	5.4	6.2	6.9	7.5	8.1	9.1
Run Bt Run Vs	140.0 5.6	144.8 6.6	153.6 6.9	165.6 7.2	169.8 7.7	171.3 8.2	168.8 9.3
True Wind	6	8	10	12	14	16	20

Fast Maneuvers

Bearing Away

Bearing away is when the boat is turned away from the wind on your current course. In anything but light and very light airs, the helmsman cannot bear away without your help in easing the mainsheet. In much windier conditions, someone also needs to ease the vang.

When you are racing, bearing away usually needs to happen when you're ducking another boat or an obstruction, or bearing away around a top mark.

As you bear away, you go through a potential power band that not only keeps your upwind speed; it enables you to accelerate hard. However, it's easy to miss the power band by bearing away too quickly or too slowly.

A useful analogy is to think of how racing drivers drive around racetracks. If the driver turns too sharply, the car skids and slows down. The 'racing line' is the smooth curve racing drivers take that enables them to keep the highest possible speed throughout the maneuver. Bearing away in sailing is the equivalent of taking the racing line but coming out of the corner as fast as or even faster than you went in. If you get it right it's like being sent out of the maneuver in a slingshot!

Every boat is different and the only way to discover the power band, and to make best use of it, is practice. Dinghy sailors simultaneously operating the helm and the mainsheet find this much easier to do than for boats with different crew on the mainsheet and helm. Dinghy sailors can easily feel the helm and the mainsheet tension together and can tell more easily how fast they should turn.

Here's the procedure for determining how fast you can turn: Sail on a beat (close-hauled) towards a buoy – this is your pretend upwind mark. As you round the buoy, the helmsman starts to turn and the mainsheet is slowly eased. The helmsman should complain that the boat is too hard to turn. You ease the mainsheet until the helmsman can change the direction without using excessive rudder. You need to note how much you eased over that period, e.g. five inches in five seconds. Now, carry on easing at the rate you previously eased. If the helmsman again complains, ease some more and note the new rate.

You should now have achieved two things: a rough idea of the rate of ease, and you and the helmsman should have a clearer idea of what you're trying to do.

Repeat the whole procedure several times. What you should now concentrate on is maximizing the boat speed through the turn. Try slower turns and faster turns until you've got the idea – until you know the target speed through the turn and the rate of ease required to achieve it. Clearly, the rate of ease will vary for every wind strength – the stronger the wind, the faster you must ease, the weaker the wind, the more slowly you must ease. You need to practice bear-away turns for all wind strengths.

Heading Up

Heading up is when the boat is steered up closer to the wind. Heading up occurs after you've born away to duck another boat, obstruction, or as you round a bottom mark.

Trimming the mainsail helps to steer the boat up into the wind, requires less rudder from the helmsman and is therefore faster.

If you're heading back up having just ducked another boat, you need to keep the sail filled and will need to trim on hard and fast. In a yacht with winches, you will need to winch the sail in.

If you're rounding a bottom mark and going from a broad reach (or run) to close-hauled, you need to get the entire mainsheet in very quickly. You need to choose the quickest method possible for your type of boat. On a dinghy, you need to take great armfuls of mainsheet. On a yacht, it's often possible to pull the mainsheet in by hand and winch the last bit. Again, this requires coordination, which is achieved through practice.

On larger boats, as the boat is steered up towards the wind, there will be tiny windows of opportunity where you can pull without the aid of a winch, and other times when you can't. You must continue to try pulling until the sail is almost in. Only then do you start to winch the sheet in.

Just like bearing away, you and the helmsman need to practice heading up. Find a buoy and nominate it as an obstruction for you to duck below, or as the bottom mark for you to round.

Tacking

Tacking is when the boat is turned through the wind from one 'tack' to the other. Good tacking depends on coordination and trim. The correct way of controlling trim during the tack for the mainsail depends on the wind strength.

Very Light Airs Tacking

- Before the tack, you are sailing in 1st gear with the traveler high and the mainsheet eased.
- As the sails collapse as the boat is turned through the wind, only one thing needs to change. The traveler needs to be set high on the new tack. Don't move the traveler until the sail has completely collapsed. There's no need to adjust the mainsheet.

Light Airs Tacking

- Before the tack, you are probably sailing in 2nd gear with the traveler high and the mainsheet trimmed on a little.
- As the sails collapse as the boat is turned through the wind, two things must be changed. The traveler needs to be set high on the new tack and you need to ease the mainsheet a little.
- After the tack completes, you are now in 1st gear and you need to get back into 2nd. As the boat accelerates, trim on the mainsheet a little.

Medium Airs Tacking

- Before the tack, you are probably sailing in 3rd or overdrive gear with the traveler around centreline and the mainsheet trimmed on.
- As the sails collapse as the boat is turned through the wind, the traveler needs to be set at the middle or lower down the track on the new tack and, if you're sailing in overdrive gear, the mainsheet needs to be eased a little.

- After the tack completes, you are now in 2nd gear and you need to get back into 3rd. As the boat accelerates, slowly pull the traveler back up the track. If you can make it back into overdrive gear, trim on the mainsheet only after the traveler is back up and the boat is going fast enough.

Stronger and Heavy Airs Tacking

In stronger airs, you will be vang sheeting (see the section on vang in Chapter 2).

- Before the tack, you should be sailing in 3rd or overdrive gear with the traveler at the bottom of the track and the mainsheet trimmed on hard
- As the sails collapse as the boat is turned through the wind, the traveler must be free to run down to the bottom of the track on the new side as the tack completes. The mainsheet needs to be eased a little
- After the tack completes, you are now in 2nd gear and you need to get back into 3rd. As the boat accelerates, trim on the mainsheet

In general, don't trim on too early – this will kill your performance and stop you from accelerating (see the section on gears earlier in this chapter).

Footing

Footing means easing the sails a couple of inches and steering lower (further away from the wind).

Footing is usually done to power through choppy water, for example through motorboat wakes.

- The cost of footing is having to point lower for a short time. The benefit is keeping speed – this is the best available compromise.

- You foot just before you hit some choppy water and as you're driving through it. If you don't foot and just keep on going, the boat will slow down dramatically and take a very long time to get back up to speed. Footing gives us the power to drive through choppy water, minimize slowing down and to accelerate again.
- Footing begins when you're given warning of impending chop. You ease the sheet a couple of inches and the helmsman steers down (away from the wind) just before the chop.
- You trim back on when you've accelerated after the chop and not before.

After you've got through the chop, an evaluation needs to be made on what gear you're in and you trim accordingly (see the section on gears).

Wind Changes

Wind changes occur for many reasons. The right course of action depends on what's happening to the wind at that particular time.

Shifts – When there's a Gust or a Lull

When the wind is under 8–10 knots, not everyone needs to be hiking-out.

- The increase in wind during a gust often manifests itself as a lift. This means that the boat is now pointing too low (too far away from the wind) and that the sail is now over trimmed. In light winds, the right course of action is for you to ease the sheet until the telltales flow evenly again. Then the helmsman points up (steers closer to the wind) helped by you as you trim the mainsail on again.

- During a lull, the decrease in wind usually appears as a header (knocks the boat off course). The sails will start to collapse and the natural reaction for the helmsman is to bear away from the wind. This is usually the wrong course of action. Instead, the helmsman should continue to steer the same course. As the boat slows down to match the new lighter wind strength, the sails will start to fill again. Since you're now sailing in a lighter wind, you will need to ease the sheet a little.

Stronger Wind

- As the wind strength builds, you should ease the traveler down and trim the mainsheet on harder to close the leech.
- If you're now over powered, you need to tighten the backstay and outhaul.
- If the stronger wind means you're sailing outside the sail and rig's wind range, consider putting in a reef.

Weaker Wind

- Weaker wind also calls for a sail trim change. As we've discussed, as the wind strength drops, you should ease the sheet to open the leech and bring up the traveler.
- If the backstay was on, you will also need to ease the backstay, outhaul and cunningham.

Sea State

Sea state challenges your ability to sail the boat at its target speed. The more waves there are and the bigger they come, the harder it will be to sail the boat in a straight line and to keep your speed.

- To make sure the sail is working as the boat is knocked around, the main should be twisted more than for flat water. This is achieved by raising the traveler and easing the sheet a little.
- Aggressive trimming and easing is required as the boat travels up and down waves. As the boat is steered up a wave, the wind at the top of the wave is stronger. This will be like sailing into a gust. You must anticipate that you will need to ease the traveler (or mainsheet in heavy airs) at the top of the wave to keep boat speed up.
- As soon as you head down the back of the wave, you must sheet back on to keep the boat powered-up and going fast. Every wave will need to be handled like a gust.

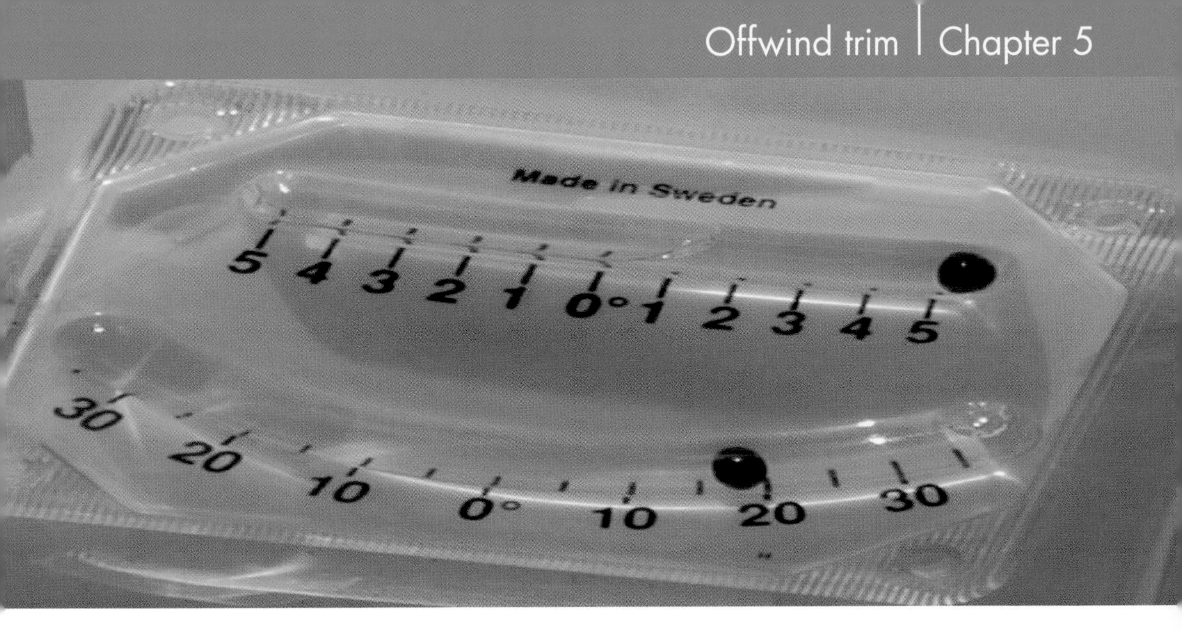

Offwind trim

Reaching

Running

Jibing

In this chapter, you will learn that trimming offwind is quite different from sailing close-hauled. Controls are used differently to make the boat go as fast as possible. We will look at close reaching, beam reaching and broad reaching. We will also illustrate how to sail downwind, including pumping and jibing – with and without spinnaker.

Reaching

Close reaching and beam reaching are busy times for the mainsail trimmer. Your responsibilities are concerned with creating the most efficient sail shape and keeping the boat on its feet. Your goal is to harness all the useful power whilst not losing control or causing the helmsman to use excessive rudder.

- The vang (not the traveler) is set to control the leech shape (twist) and should be set according to the instructions in Chapter 2.
- The mainsheet – not the traveler – is used to keep the boat on its feet and going fast.

If you have instruments and target boat speeds, you should use these to tell how well you're doing.

You must be prepared to sheet out as soon as the boat becomes over powered. Similarly, you need to get the sheet back on fast enough to keep the boat speed up.

It's essential for the helmsman to communicate when he is over powered. The objective is for you to acquire knowledge based on the heel of the boat, your feel for the wind, and the accumulated feedback from the helmsman.

Sheeting out too fast loses power. In stronger winds, sheeting out too slowly causes the helmsman to slow the boat by using too much rudder and, worse still, can cause the boat to round-up into the wind!

Emergency Vang!

Sailing in bigger winds makes the boat more challenging to keep under control. Sometimes, simply easing the mainsheet isn't enough as you're hit by a gust. If you're in the process of being caught out and the boat is going out of control, the last resort is to release the vang. This has the effect of de-powering the mainsail so that the helmsman can steer the desired course, rather than the boat being rounded-up into the wind. It's a good idea for one person to be nominated to release the vang if it's necessary to do so. It's better still if that person can operate the vang from the high side of the boat. This ensures that their weight doesn't go to the low side of the boat to operate the vang, and hence make matters even worse!

Running

The game changes considerably down wind. The mainsail is no longer acting as an airfoil. Instead, the main is acting like an air dam. Therefore, all you need to do is set the vang so that the top batten of the mainsail is parallel to the boom.

There are two schools of thought about whether you ease or tighten the outhaul downwind. One school states that you ease the outhaul to maximize the belly (and hence power) of the sail. The other school states that easing the outhaul reduces the sail area and that since the sail is not acting as an airfoil, and since you must therefore maximize the sail area, you must set the outhaul tight.

Given the fact that championship sailors subscribe to both theories in apparently equal measure, we suggest that it probably doesn't matter whether you have a trimmed or eased outhaul down wind.

Pumping

Although sailing downwind is usually a relatively quiet time for the mainsail trimmer, there are some things you need to be prepared to do at a moment's notice.

There's a huge amount of energy in waves that you can use to propel the boat forward. Every wave is an opportunity to surf. *The Racing Rules of Sailing*[1] allow you to 'pump' the mainsail once for every wave. This is important. It means that every time a wave approaches (from behind) you can pump the main to get an extra spurt of speed in the hope of catching the wave. This is the equivalent of what surfers spend all their time doing.

Again, the technique requires some coordination with the helmsman. As the wave approaches, the helmsman steers up (closer to the wind) and you trim on the mainsheet to accelerate. Just before the wave reaches the boat, you ease the main out again at same time as the helmsman steers downwind again. As the back of the boat becomes raised by the wave, you pump the mainsail.

To pump the mainsail, grab the whole sheet (between the boom and the cockpit) and pull it. Don't winch the sheet or grab the normal working end of the sheet. Grab the rope that's attached to the boom.

If you get it right, the boat will be picked up by the wave and the boat will be driven forward much faster than by wind alone.

---

[1] The ISAF (International Sailing Federation) Racing Rules of Sailing are revised and published every four years. Printed copies are available through your member National Authority or can be purchased from the ISAF Secretariat using the publication order form.

Pumping

Mainsheet in hand

Looking backwards towards approaching wave - ready to pump

Jibing

Jibing is when the boat is turned through the wind whilst the wind is behind the boat.

For many people, jibing is one of the most frightening times sailing on a boat. This is primarily because the boom moves all the way from one side of the boat to the other.

Anything in the way is likely to be struck hard. Hospitals near marinas are very used to dealing with weekend sailors with head injuries!

Your jibing technique will depend on whether you're sailing under spinnaker and, if so, how fast you can execute the jibe.

Jibing under Symmetric Spinnaker

If you're sailing under symmetric spinnaker, it is possible (and desirable) to keep the spinnaker filled during the jibe. Doing so ensures that boat speed is retained and that the apparent wind behind the mainsail is relatively light. This makes jibing under spinnaker a lot easier and safer for anyone in the cockpit.

There are two styles of jibing under symmetric spinnaker depending on how fast you can jibe the spinnaker itself.

Fast Jibes

If the crew can execute fast jibes with the spinnaker, the mainsail can be pulled across at the same time as the spinnaker, in one continuous movement.

On smaller boats, you can literally throw the mainsheet from one side to the other by grabbing the sheet directly between the boom and the cockpit.

On larger boats, where the mainsheet runs back to the mast, it pays to have someone at the mast pulling the mainsheet in and releasing it as soon as the boom goes through the wind. That person needs to be very careful not to get their fingers caught in a block as the mainsheet runs through.

Slow Jibes

If the crew cannot execute a fast jibe with the spinnaker, it's still imperative to keep the spinnaker filled throughout the jibe. The best way of achieving this is for the boom to be brought in to the centreline of the boat, thus making it much easier to keep the spinnaker filled during the long jibe (as the centered boom does not blanket the spinnaker). As soon as the spinnaker has been successfully jibed, the boom can be eased out on the new side.

Jibing under Asymmetric Spinnaker

The mainsail is used to blanket the asymmetric spinnaker while the jibe completes. This gives the spinnaker trimmer the opportunity of working with a de-powered sail as the clew is pulled around the front of the boat. The jibe works as follows:

As the boat is prepared for the jibe, the helm steers the boat away from the wind onto a run. Correspondingly, you need to ease the mainsheet so that the boom is all the way out. This ensures that the spinnaker is now behind the mainsail with respect to the wind. Next, the spinnaker is pulled around the windward side of the boat. When the bulk of the spinnaker is around onto the new side (someone should call this if you can't see the spinnaker) it's time to jibe the main quickly across. The last of the spinnaker is pulled around, the helm steers up to fill the spinnaker and you need to trim the main on quickly to the new reach.

The faster this all happens, the less apparent wind there is working against you and the spinnaker trimmer during the jibe. Also, faster jibes lose less speed and distance.

Beware jibing the main before the majority of spinnaker is around the boat onto the new side (this advice goes for the helmsman too). Jibing too quickly is likely to cause the sail to fill on the wrong side of the boat and then wrap itself around the forestay. The ensuing tangle can quickly develop into nightmarish proportions and become very hard to recover from!

Jibing without a Spinnaker

Jibing without a spinnaker is more frightening and hazardous for those in the cockpit. If the jibe takes too long, the boat will slow down and the apparent wind will increase, thus causing the boom to come across the boat with even more dangerous ferocity.

The key to executing a mainsail jibe (without the benefit of a spinnaker maintaining boat speed) is to complete the maneuver as quickly as possible.

On larger boats and in heavier airs, it pays to pull the mainsheet in before the jibe to reduce the distance the boom can travel during the jibe. The mainsheet is then immediately eased again as the jibe completes. This minimizes the risk of damage to boom and mast.

Hoisting, dropping and reefing

Hoisting

Dropping

Reefing

Hoisting and dropping the mainsail normally requires that the boat be facing into the wind (head to wind). This ensures that the sail can move freely up and down the mast

Hoisting

There are four points to remember when you hoist the mainsail:

First, make sure the vang, mainsheet and outhaul are all loose before you start hoisting.

Second, make sure the backstay is off. If the sail is hoisted while the backstay is on, at some point later when the backstay is eased, there will be too much halyard tension.

Third, don't over-tighten the halyard as you complete the hoist. Although you need to achieve a full hoist, you must avoid excessive luff tension. The best way of ensuring this is to have the person doing the hoisting, e.g. at the mast or in the cockpit, looking at the luff and the belly as the sail goes up. The goal is to have the sail fully hoisted and for the belly to be in the middle of the sail. This should happen as the horizontal wrinkles at the luff have just disappeared.

Fourth, make sure that the hoist is completed fast, that the boat bears away, and that the sail is full as soon as possible. Flogging is a sure way to destroy sail shape and hence the sail's performance.

Dropping

There are three important points to remember when you drop the main:

First, make sure the main halyard is ready to run before you start the drop. The goal is to get the sail down as fast as possible with as little flogging as possible. If the halyard is caught as the sail comes down, the sail may be flogged for a couple of minutes before the tangle is resolved.

Second, ensure that the outhaul is eased a little before you start the drop. This prevents tension in the sail's foot from stopping the sail running down the mast.

Third, if the boat normally has the mainsail flaked onto the boom and the boat doesn't have lazy jacks, don't flake the sail as you drop – unless there's no wind. It's much faster to get the sail onto the deck before you flake and therefore avoid any unnecessary flogging.

Reefing

Reefing is performed if the boat is still over powered after the sail has been de-powered using the backstay and outhaul.

Reefing involves reducing whole chunks of sail area. This enables an acceptable balance between lift and drag to be restored. If you don't reef and you're sailing much outside the rig and sail's wind range, the forces of drag will prevent you from being able to achieve the highest possible boat speeds and control, including pointing.

If there is any chance you're going to need to reef, you must ensure the reefing equipment is ready to use and that you know how it operates.

Reefs can be taken out or put in while you're sailing but it requires good coordination and practice.

Here's the procedure for putting in a single reef under sail:

- Have someone ready to ease the main halyard gradually.
- Have someone attach the cunningham onto the reef point.
- Have someone ready to wind on the reefing line.

When everyone's ready, the three following things happen at once – if necessary, the boat is briefly luffed for the operation:

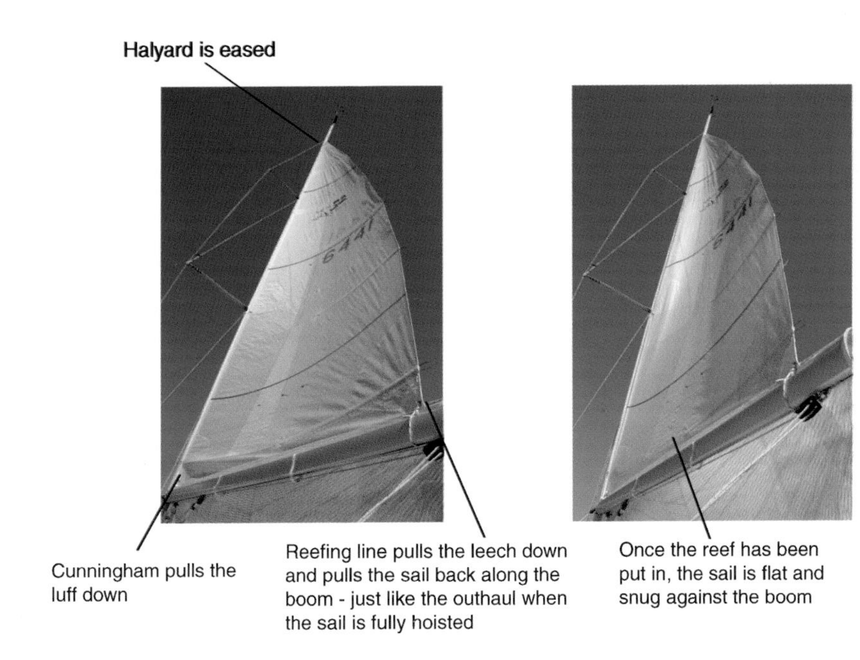

1. halyard is eased;
2. reefing line is tightened;
3. cunningham is pulled on.

The aim is to ease the sail down until the reef points are just at the boom – but no lower. There must be no wrinkles in the luff and the foot should be flat.
The procedure for taking out the single reef is the same but in reverse:

- Have someone ready to wind on the main halyard gradually.
- Have someone ready to ease the cunningham.
- Have someone ready to ease the reefing line.

When everyone's ready, the three following things happen at once – if necessary, the boat is briefly luffed for the operation:

1. halyard is wound on;
2. reefing line is eased;
3. cunningham is eased.

Conclusion

Throughout this book we've talked about, and illustrated, the key concepts of mainsail trimming. We've explained the core principles of lift and drag, the mainsail and its controls, the complete mainsail shapes and how to achieve them, and we've looked in detail at how to sail fast. We've encouraged you to see shapes rather than learning figures, explained how important speed is to acquire and how to achieve it, and we've avoided jargon and focused only on what you need to know.
In short, good mainsail trimming requires that you:

- understand what shape you need for the current conditions, and how to achieve it
- focus on accelerating to and keeping your target speeds
- keep your heel – never loose it
- communicate with the helmsman, the jib trimmer and, if present, the tactician.

That's it, good luck, but if you've followed closely, you won't need it!

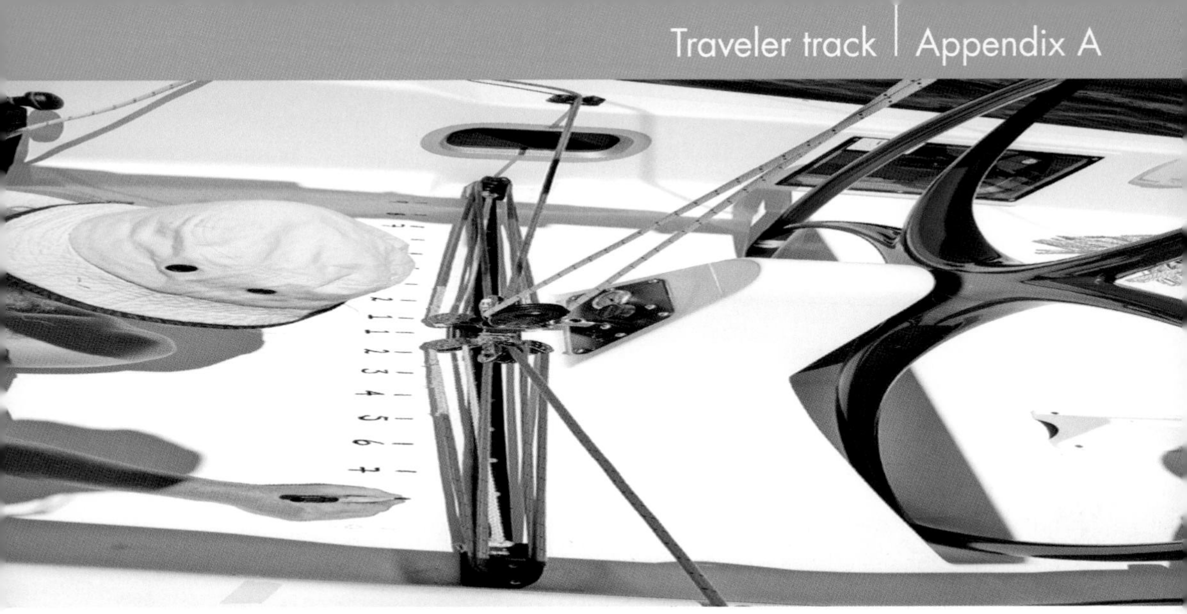

Appendix A – Traveler track

Setting up the traveler track is the key to achieving fast settings consistently.

The procedure is as follows:

Using an indelible marker, and starting from the middle of the track, write numbers starting at 0 alongside the track. The numbers should go up in measures of the length of the traveler car until you reach the end of the track.

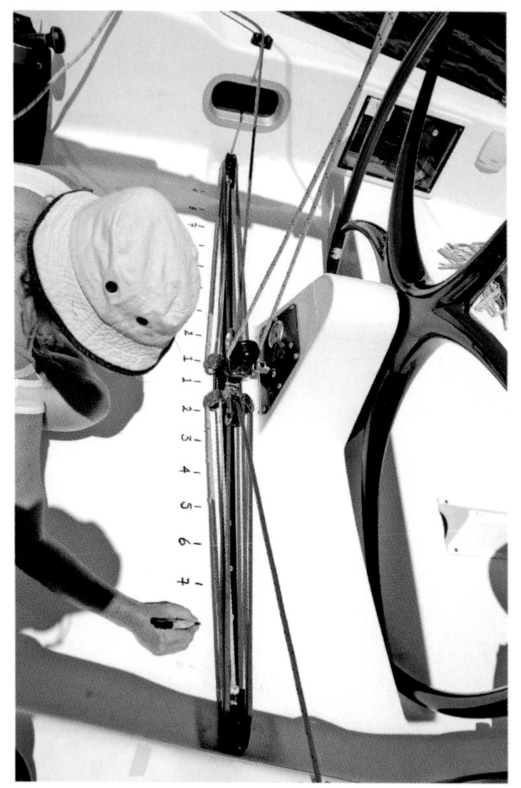

Next, draw marks halfway between the numbers.

You now have the basis for recording and repeating fast settings for the traveler for each wind strength.

Appendix B – Trim table template

Wind speed	**0–5 knots Very Light**	**5–8 knots Light**	**8–16 knots Medium**	**16–25 knots Strong**	**25 knots + Heavy**
Backstay	Half on	Off	Off → On	On	On
Outhaul	Half on	eased	Eased → On	On	On
Traveler	Up	Up	Upper middle → Lower middle	Down	Middle (Reefed)
Mainsheet	Eased	Mostly eased	Mostly eased → firm	Firm → Tight	Tight (Reefed)
Boom	Just above center line	At center line	At center line	Just below center line	Below center line

Appendix C – Points of sail

Wind

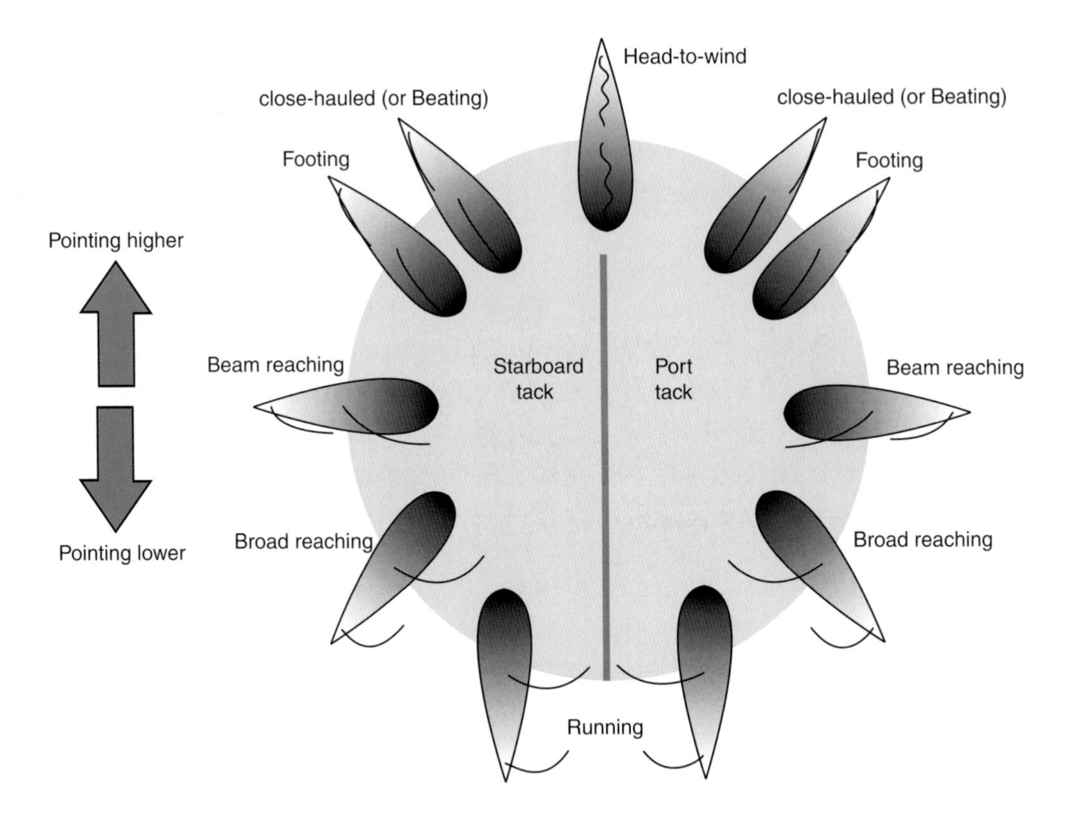
Head-to-wind
close-hauled (or Beating)
close-hauled (or Beating)
Footing
Footing
Pointing higher
Beam reaching
Starboard tack
Port tack
Beam reaching
Broad reaching
Broad reaching
Pointing lower
Running

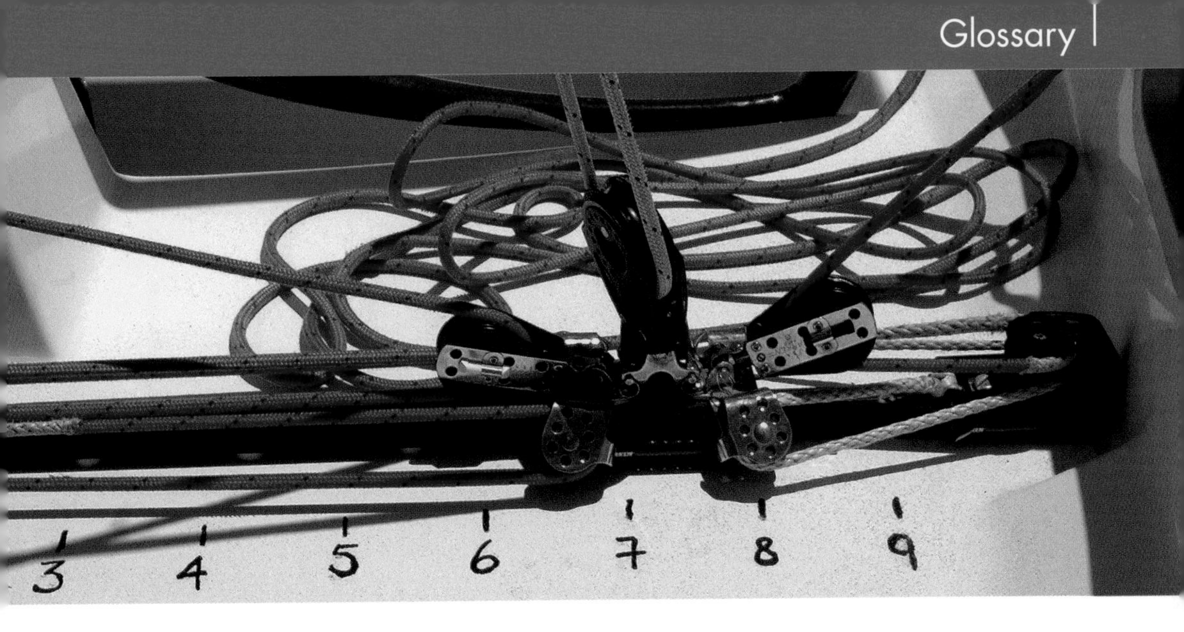

Glossary

Airs	A way of referring to wind strength, e.g. very light airs, light airs, medium airs, strong airs, heavy airs.
Airfoil	A shape used to create lift, e.g. a sail or an airplane wing. Also known as an aerofoil.
Apparent wind	The wind experienced on a boat. If a boat is stationary, the apparent wind is the same as the true wind. If the boat is moving towards the wind, the apparent wind is greater than the true wind. If you boat is moving away from the wind, the apparent wind is less than the true wind.
Belly	The fattest part of a sail also known as the draft.
Beating	Sailing as close to the wind as possible, also known as sailing close-hauled.
Back	Towards the back (stern) of the boat.
Backstay	The wire running from the top of the mast to the back of the boat, which controls forestay tension (jib power) and mainsail depth in the upper two thirds of the sail.
Block	A pulley.
Bow	The forward most part of a boat's hull.
Broad reaching	Sailing with the wind almost behind you (see Appendix C – Points of Sail).
Cleat	A device for making a rope fast, i.e. holding it and not letting it go despite the rope being under load.
Clew	Corner of sail that a sheet is connected to.

Close-hauled	Sailing as close as possible to the wind (also known as beating).
Closed	Typically referring to the top of mainsail's leech and meaning little or no twist.
Cunningham	An assembly that connects to the luff towards the bottom of the sail to apply down force to tighten the luff. For a boat with a backstay, its primary purpose is to take up the slack when the backstay is tensioned. For smaller boats without backstays, the cunningham is used to bring the belly of the sail forward again after high mainsheet tension has caused the belly to move back.
Depth	The size of the sail's draft, also known as belly. More depth equals more power.
Down	A term used to indicate something further away from the wind (as opposed to up).
Drag	The side effect of lift that causes the boat to heel over and slow down.
Ease	To sheet out or to loosen a rope, typically a sheet.
Foot	The bottom edge of the sail.
Gust	A temporary increase in the wind.
Halyard	A rope connected to the head (top) of a sail used to hoist the sai.l
Foot (1)	To turn slightly away from the wind while easing the sails, typically to power-up through chop.
Foot (2)	The bottom edge of the sail.
Footing	See Foot (1).
Forward	Towards the front (bow) of the boat.
Header	When the wind changes direction disadvantageously and forces you to sail further away from your course than the old wind allowed. Also known as a knock.
Heel	The angle that a boat assumes as it is pushed over, whether by the wind or crew weight, or any other force.
Helmsman	The person steering the boat, whether by tiller or by wheel.
Jib	The front sail used on a yacht or a dinghy. Often referred to as a headsail.

Leech	The edge of the sail furthest back from the wind.
Lift (1)	The useful force that an airfoil (such as a sail) creates.
Lift (2)	When the wind changes direction advantageously and allows you to sail closer to your course than the old wind allowed.
Luff (1)	The leading edge of a sail attached to the forestay.
Luff (2)	To turn the boat into the wind so that the sails start to collapse.
Luffing (1)	Easing the sails so that they are no longer completely filled.
Luffing (2)	Turning the boat into the wind so that the sails are no longer completely filled.
Lull	A temporary drop in the wind.
New	A term used to distinguish sheets on one side of the boat from the other. As you tack or jibe, the new sheet will become the working sheet after the tack or jibe. The old sheet is the working sheet before the tack or jibe. Can also be used to refer to the wind before and after a wind shift.
Old	See New above.
Open	Typically referring to the top of the sail's leech and meaning that the sail has twist.
Point	A term used to say how far away from the wind you're sailing.
Point (2)	A demand such as 'please point down a bit', meaning please steer away from the wind a little.
Pointing down	Steering further away from the wind.
Pointing up	Steering closer to the wind.
Port	The left hand side of the boat as you look forward towards the front (bow) of the boat.
Rudder	The submerged part of the steering assembly.
Running	Sailing with the wind almost or actually behind you.
Sail	One of the following: headsail (either jib or genoa), mainsail or spinnaker.
Sheet	A rope connected to a sail used for trim. This is our primary control rope for the mainsail.
Sheet on	The action of tightening or trimming-on the sheet.

Sheet out	The action of loosening or easing the sheet.
Shift	When the wind changes direction.
Shrouds	The wires either side of the mast that hold the mast in place and control its overall shape.
Spreaders	The struts or spars that stick out from the mast and connect to the shrouds to hold the mast up and to control rig tension.
Stern	The back of a boat's hull.
Starboard	The right hand side of the boat as you look forward towards the front (bow) of the boat.
Tack (1)	The front and bottom corner of the sail which is attached to the boat.
Tack (2)	The current tack you're on, i.e. starboard tack (wind coming over the starboard side) or port tack (wind coming from the port side).
Tack (3)	See Tacking.
Tacking	Changing course by steering the boat through the wind when the wind is in front of the boat.
Telltale	The lightweight thread or wool attached to the sails used to visualize the wind's presence around a sail. If they are different colors, then the red one is the port telltale and the green one is starboard.
Tiller	The handle connected to the rudder used to steer the boat.
Trim	The overall state of a sail, i.e. how it's set by its controls such as the halyard, cunningham, sheet, vang, outhaul or traveler.
Trim	To sheet on or tighten a rope, typically a sheet.
Trimmer	The person whose responsibility it is to control the set of the sail.
Trimming	The activity of controlling the set of a sail.
Twist	The amount that the top of the leech is angled out from the boat compared to the leech at the bottom of the sail.
Up	A term used to indicate something closer to the wind (as opposed to down).
Wheel	The wheel connected to the rudder used to steer the boat.
Winch	A drum that is used to make pulling on ropes easier. It's normally operated with a removable winch handle.

Index

Airs, 18, 21, 25, 26, 28, 36, 38, 53, 57, 59
Airfoil, 2, 3, 4, 31, 38, 67

Belly, 8, 9, 26, 27, 28, 29, 30
Beating, 14, 86, 88
Backing, 6, 19, 27, 54, 55, 58, 59, 60, 68
Backstay, 6, 26, 27, 28, 30, 31, 32, 36
Block, 12, 31, 70, 88
Bow, 55, 88, 89, 90, 91
Broad reach, 15, 26, 58, 88

Car, 45, 57, 82
Cleat, 88
Clew, 10, 23, 29, 32, 37, 71
Close-hauled, 14, 16, 19, 22, 38, 49, 86
Cunningham, 6, 11, 26, 28, 77, 89

Depth, 1, 8, 9, 89
Drag, 1, 2, 5, 8, 76, 80

Fast, 15, 22, 43, 44, 52, 57, 70, 82
Foot, 10, 28, 29, 36, 60, 61, 77
Footing, 44, 49, 60, 61, 86
Forestay, 7, 71, 88

Gears, 12, 43, 45, 47
Gust, 15, 55, 61, 89

Halyard, 11, 26, 74, 76, 89, 91
Head, 10, 23, 58
Header, 62, 89
Heel, 5, 15, 18, 19, 21, 43, 44, 50, 51
Helmsman, 46, 53, 57, 58, 68

Inclinometer, 51

Leech, 9, 16
Leech telltales, 39, 40
Lift, 2, 61
Luff, 90
Lull, 61, 62

Outhaul, 7, 28, 29, 67, 76

Points of Sail, 14, 24, 56
Pointing, 47, 48
Port, 90

Reef, 9, 76
Rudder, 5, 52
Running, 16, 25, 67

Sheet, 68, 90
Shift, 61, 91
Shrouds, 91

Stern, 91
Starboard, 91

Tack, 27, 46
Tacking, 59, 60
Telltale, 38, 39, 40, 41, 91
Tiller, 52, 91
Traveler, 9, 16, 17, 21, 22, 24, 59
Twist, 9, 25, 91

Winch, 58, 91